endorsed for
edexcel

REVISE EDEXCEL GCSE

Geography B
Evolving Planet
For the linear specification first teaching 2012

D1634505

REVISION WORKBOOK

Series Consultant: Harry Smith

Author: David Holmes

A note from the publisher

In order to ensure that this resource offers high-quality support for the associated Edexcel qualification, it has been through a review process by the awarding body to confirm that it fully covers the teaching and learning content of the specification or part of a specification at which it is aimed, and demonstrates an appropriate balance between the development of subject skills, knowledge and understanding, in addition to preparation for assessment.

While the publishers have made every attempt to ensure that advice on the qualification and its assessment is accurate, the official specification and associated assessment guidance materials are the only authoritative source of information and should always be referred to for definitive guidance.

Edexcel examiners have not contributed to any sections in this resource relevant to examination papers for which they have responsibility.

No material from an endorsed resource will be used verbatim in any assessment set by Edexcel.

Endorsement of a resource does not mean that the resource is required to achieve this Edexcel qualification, nor does it mean that it is the only suitable material available to support the qualification, and any resource lists produced by the awarding body shall include this and other appropriate resources.

Contents

Make sure you know which topics you have studied.

A small bit of small print
Edexcel publishes Sample Assessment Material and the Specification on its website. This is the official content and this book should be used in conjunction with it. The questions in *Now try this* have been written to help you practise every topic in the book. Remember: the real exam questions may not look like this.

Target grade ranges
Target grade ranges are quoted in this book for some of the questions. Students targeting this grade range should be aiming to get most of the marks available. Students targeting a higher grade should be aiming to get all the marks available.

Moving tectonic plates

FOUNDN
G/E

1 Study Figure 1.

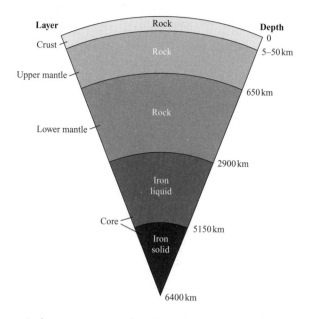

Figure 1 A cross section showing the structure of the Earth.

What is the thickest layer in the cross section of the Earth?   **(1 mark)**

☐ **A** Crust

☐ **B** Upper mantle

☑ **C** Lower mantle

☐ **D** Core

HIGHER
C/B

2 State **one difference** between oceanic and continental crust.   **(2 marks)**

Guided

EXAM ALERT

Continental crust is normally made of .rocks .that .cooled .below .the .surface whereas

.oceanic .crust .....is .much .thinner .& .made. .up .of .rocks .such

.as .basalt

....

....

> Exam questions similar to this have proved tricky – be prepared! **ResultsPlus**

HIGHER
B

3 Outline what is meant by the idea of a **convection current** in relation to plate tectonics.   **(2 marks)**

.Convection .currents ...is .the .movement .of .mantle .material

.due .to .heating .and .cooling .resulting ...in .the .mantle

.material .rising .and .sinking.

....

> Outline means you have to describe the main features of something and give some development. But, don't waste time explaining or examining here.

Plate boundaries, volcanoes and earthquakes

HIGHER
C/B

1   Study Figure 1.

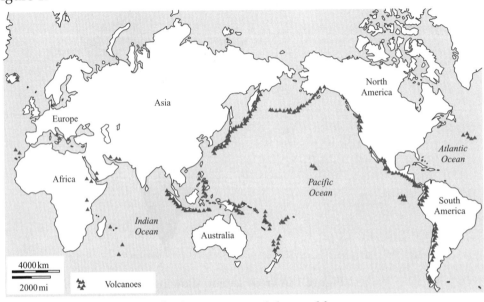

Figure 1 The distribution of volcanoes around the world.

State **two** facts about the distribution of volcanoes in Figure 1.   **(2 marks)**

1   ...mainly along coast lines ledge of continents...

2   ...around / surrounding pacific ocean...

FOUNDN
D

2   Describe the characteristics of **one** type of plate boundary.   **(2 marks)**

Type of plate boundary: ...conservative plate boundry.........................
plates slide past eachother. No crust formed or
destroyed & no volcanic formation. Great strain builds up
along junction with sudden luches at fault. Earthquakes frequent

> In this case, 'characteristics'
> basically means 'what happens'.

FOUNDN
C

> Guided

3   Explain why earthquakes are usually found close to **destructive** plate boundaries.   **(4 marks)**

Earthquakes are associated with destructive plate boundaries which are sites of subduction.
As plates move together, this creates immense heat and pressure which in turn ...makes......
the denser basaltic oceanic plate sink below continental
plate. Known as subduction. Creates deep ocean trench.
Oceanic plate is subjected to increased pressure & temp.
Cause light weight material to melt & rise to the surfau
to form volcanoes. collision of plates lifts & buckies to
also create fold mantains.

Volcanic and earthquake hazards

FOUNDN
D

1 Give **two** reasons why some earthquakes kill more people than others. **(2 marks)**

1 Size of event / capacity of p.p (how prepared)

2 the vunerability of the population - poverty & high density increase vunerability

HIGHER
B

2 Outline the main factors that may control the **strength** and **size** of a tsunami. **(3 marks)**

....

....

....

....

....

....

> Always read the question carefully –
> this is asking about physical factors,
> so you should not include ideas about
> vulnerability of the population!

HIGHER
B

3 Study Figure 1.

Figure 1 The 921 Earthquake Museum in Taiwan which shows the entrance to the Guangfu Junior School that was destroyed in 1999.

Explain **one** impact of the earthquake shown in Figure 1. **(2 marks)**

Guided The photo shows that one impact of the earthquake was ....

....

....

....

Managing earthquake and volcanic hazards

FOUNDN

E

1 Describe what is meant by the term 'long-term relief'. **(2 marks)**

....

....

....

....

HIGHER

B

2 Outline **two** actions that can be taken to reduce the impact of earthquakes. **(4 marks)**

....

....

....

....

....

....

....

....

....

HIGHER

A

3 Describe how hazard-resistant design can be used in different places to reduce the impact of tectonic hazards. **(6 marks)**

⟩ **Guided** ⟩

Japan suffers earthquake and tsunami hazards due to its location close to the Pacific Plate boundary. Aseismic buildings are common in Tokyo – they reduce the risk of building collapse even during 7–8 magnitude events. They are specially made of reinforced materials and can withstand extreme shaking. ....

....

....

....

....

....

....

....

Make sure the quality of your written communication is also good.

This is a 6 mark question. It's a good idea to include details and facts wherever possible. Here, it would be sensible to contrast Japan with a less developed country ('different places' in the question).

Earthquake case study

FOUNDN
E

1 Outline what is meant by the 'primary impact' of an earthquake.   **(2 marks)**

....

....

FOUNDN
C

Guided

2 Describe the impacts of **one** named earthquake on property and people.   **(4 marks)**

Named earthquake event: Haiti, 2010

The 2010 Haiti earthquake impacted on people in lots of ways. Around 300 000 people were

killed and many more were injured. Also ....

....

....

....

....

....

HIGHER
A

3 For a named earthquake, compare the primary and secondary impacts.   **(6 marks)**

Named earthquake event:

....

....

....

....

....

....

....

....

....

....

....

....

This question isn't just asking you
what the primary and secondary
impacts were, but to **compare**
them so remember to do that!

Volcanic eruptions

FOUNDN

E

1 Identify **two** threats from volcanic eruptions. **(2 marks)**

1 ....

2 ....

HIGHER

B

Guided

2 Describe the economic and social impacts of a named volcanic event. **(4 marks)**

Named volcanic event: Eyjafjallajökull, Iceland

This event in 2010 had a limited economic and social impact on the island itself (it was in the south and winds blew the ash cloud away from the capital), but unusually it had major international effects. This was mostly linked to air travel and the stopping of flights over much of northern Europe. ....

....

....

....

....

HIGHER

A

3 For a named volcanic event, explain the role of prediction and warning in order to reduce the impacts. **(6 marks)**

Named volcanic event:

....

....

....

....

....

....

....

....

....

....

....

....

> There are two different aspects being asked for here – prediction and warning – so make sure you deal with both in your answer.

Past climate change

FOUNDN

G

1 Study Figure 1.

(1 mark)

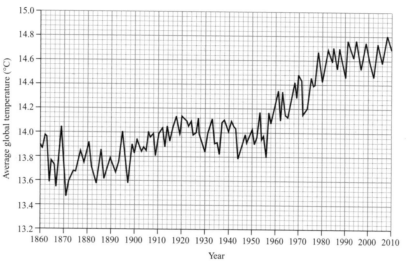

Figure 1 Average global temperature, 1860–2010

Which **one** of the following statements is correct?

☐ **A** Average global temperature has fallen since 1860.

☐ **B** Average global temperature was higher in 2010 than in 1860.

☐ **C** Average global temperature has remained the same since 1860.

☐ **D** Average global temperature was lower in 2010 than in 1860.

HIGHER

C

2 Using Figure 1, summarise the changes in average global temperature between 1860 and 2010.

(2 marks)

Guided

Overall, the graph shows a gradual increase in global temperatures from a mean of ....

....

....

....

> Make sure you use numbers and data from Figure 1 to develop your answer. Words such as 'rising', 'falling', 'trend', 'change', 'increase' and 'pattern' may also be useful to use in your answer.

The impact of climate change

HIGHER

B

Guided

EXAM ALERT

1 Outline **two** impacts of **past** climate change on the UK. **(4 marks)**

Impact 1: When climate was cooler during the Little Ice Age (about 300–400 years ago), there is evidence of the River Thames freezing. The impact of this would have been much colder average temperatures and it would have reduced the range and type of crops grown in the UK.

Impact 2: ....

....

....

....

....

> Exam questions similar to this have proved tricky – be prepared! **ResultsPlus**

FOUNDN

E

2 State **two** natural causes of climate change in the past. **(2 marks)**

1 ....

....

2 ....

....

FOUNDN

D/C

3 Using examples, describe how past climate change has affected both **people** and **ecosystems**. **(4 marks)**

Effects on people: ....

....

....

....

Effects on ecosystems: ....

....

....

....

> To do well with questions that specify two things, make sure you answer both parts (people **and** ecosystems).

Present and future climate change

1 Study Figure 1.

Figure 1 A coal-fired power station in Shropshire, UK.

Which **one** of the following is a **human** activity probably linked to changes in global temperatures? **(1 mark)**

☐ **A** Emissions of greenhouse gases from volcanoes.

☐ **B** Generating energy from renewable sources such as wind and solar.

☐ **C** Combustion of fossil fuels to power vehicles, e.g. cars and trains.

☐ **D** Changes in the Earth's orbit around the Sun.

2 Describe **one** human activity that is thought to have led to an increase in greenhouse gases. **(2 marks)**

....

....

....

....

3 Apart from the burning of coal in power stations, give **two** other types of human activity which are thought to contribute to the **enhanced greenhouse effect**. **(4 marks)**

1 Farming intensification has meant that ....

....

....

....

2 ....

....

....

....

> Don't just say 'coal-fired power station', i.e. a lift from the resource. You must try and link burning of fossil fuels to changes in atmospheric concentration of certain gases, e.g. carbon dioxide.

Climate change challenges

HIGHER

C/B

1 Study Figure 1.

Describe **one** possible impact of more frequent flooding as a result of climate change in the UK. **(2 marks)**

....

....

....

....

....

....

> In the exam paper, each new question will have some stimulus at the beginning – this could be a photo or illustration, a piece of text or some data. Whatever it is, it is there to help you answer the questions so make sure you look at it properly!

Figure 1 A pub which has been flooded by a river in the UK.

FOUNDN

C

2 For a named **developing** country, explain how climate change may affect the people who live there. **(4 marks)**

> Guided

Named county: Bangladesh

Bangladesh is a large and very low-lying country with a big coastline. As such, its population is vulnerable to even a small (20–40 cm) sea level rise from climate change. This would bring repeated coastal flooding and loss of coastal communities. A warming world may also bring other challenges for Bangladesh. ....

....

....

....

....

....

> Try to give reasons and examples to support your points. Make sure you focus on the impact on people.

Climate change in the UK

FOUNDN
C

1   For the UK, describe **one positive** and **one negative** economic impact of higher global temperatures. **(4 marks)**

Positive impact:

> **Guided**

Parts of the UK will become warmer, increasing the length of the growing season for farmers. This may mean that a greater range of crops can be grown in some areas, as well as an increase in the yield (output) per hectare.

Negative impact:

....

....

....

....

HIGHER
A

2   Explain a range of possible **environmental challenges** that could occur as a result of future climate change on the UK. **(6 marks)**

....

....

....

....

....

....

....

....

....

....

....

....

....

> The command for this 6 mark question is 'explain'. That means that you have to make a statement and then give some reasons for that idea, e.g. sea level rise will lead to a more intense erosion of the coastline as it will be less protected from severe storms (associated with climate change). Think about the possible impacts of climate change.

Climate change in Bangladesh

1   Study Figure 1.

> **Climate change in Bangladesh**
>
> Food production will be particularly sensitive to climate change, because crop yields depend directly on climatic conditions (temperature and rainfall patterns) and could lead to food yields being reduced by as much as 30%. Tropical cyclones will become stronger, with faster wind speeds increasing the amount of damage they cause; floods will become more common due to changing rainfall patterns and glacier melt in the summer; sea level rise could cover low-lying areas; and the changing climate may indirectly cause misery by increasing the amount of disease and conflict between people.

Figure 1 Some of the predicted range of impacts from climate change on Bangladesh by 2100.

State **two** of the predicted impacts of climate change on **people** in Bangladesh.   **(2 marks)**

> **Guided**

1   People may suffer from more disease.

2   ....

....

HIGHER

B/A

> **Guided**

2   Using Figure 1 and your own knowledge, outline the range of impacts that climate change might bring to the people of Bangladesh.   **(4 marks)**

There are a range of impacts from climate change which will affect Bangladesh and its people. One of the most serious of these is the predicted drop in crop yields. This could lead to hunger, malnutrition and in worse cases starvation. It adds increased stress to an already poor population, especially if yields drop by more than the estimated 30 per cent.

....

....

....

....

....

....

> The 'outline' command word in this context is directing you to be reflective in terms of what you are writing. There is also a focus on 'people' which should be recognised. You should try to use the information in Figure 1 and then expand it further using some geographical thinking. See the Guided part of the answer to help get you started.

Distribution of biomes

1   Study Figure 1.

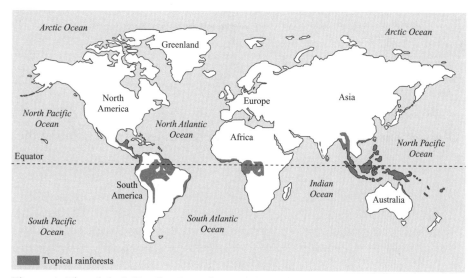

Figure 1 The global distribution of tropical rainforests.

Which **one** of the following statements is correct?   **(1 mark)**

☐ **A**   There are many tropical rainforests in the northern hemisphere.

☐ **B**   Tropical rainforests are found all over the world.

☐ **C**   Tropical rainforests tend to be found in a restricted band near the Equator.

☐ **D**   Tropical rainforests are found only in South America and Africa.

2   Describe the global distribution of tropical rainforests shown in Figure 1.   **(2 marks)**

....

....

....

....

3   Describe how climate and other factors can affect biomes.   **(4 marks)**

Guided

Climate can control the distribution of biomes through variations in temperature and precipitation, which limit the distribution (range) of plants and animals.

> Make sure you write about local factors (altitude, geology and aspect for instance). You can also write about the influence of people and link it to an example / place.

There are many local factors that can influence

biomes ....

....

....

....

....

....

A life-support system

FOUNDN
E/D

1 State **two** ways in which the biosphere acts like a life-support system. **(2 marks)**

1 ....

....

2 ....

....

FOUNDN
E/D

2 Describe **one** function of a tropical rainforest in terms of regulating the atmosphere.

(2 marks)

....

....

....

....

HIGHER
C

3 Timber is an example of an ecosystem 'good'. State **two** other examples of ecosystem **goods**. **(2 marks)**

Good 1: ....

....

Good 2: ....

....

HIGHER
A

4 Using examples, explain the importance of the biosphere for providing people with a number of different goods. **(6 marks)**

Guided

Tropical rainforests, for example, are important in providing a range of different timber goods (especially hardwoods) that people use for building houses and making furniture. ....

....

....

....

....

....

....

....

....

....

....

'Examples' can refer to examples of places / locations or simply different types of goods and services.

Threats to the biosphere

FOUNDN
E/D

1 Outline **one** way in which people can damage the biosphere.   **(2 marks)**

....

....

....

....

FOUNDN
D/C

2 Describe **two** possible effects of deforestation (cutting down trees) on the environment.

     (4 marks)

▸ **Guided** ▸

Effect 1: Locally, deforestation can lead to problems associated with soil erosion which in turn reduces the fertility and productivity of an area.

Effect 2: ....

....

....

....

HIGHER
A

3 Using example(s), explain how people may be causing **direct damage** to a biosphere or ecosystem.   **(6 marks)**

....

....

....

....

....

....

....

....

....

....

....

> The focus of the question is direct damage (often at a local scale). Causes may include tourism, overfishing, deforestation, mining, etc. Climate change (indirect action) would not be relevant here.

Management of the biosphere

1 Give **one** example of a **player** (organisation or individual) involved in conservation of the biosphere. **(1 mark)**

....

....

2 Study Figure 1.

THE NATIONAL TRUST
OPEN TO THE PUBLIC
(SUBJECT TO THE BYELAWS
ON THE BACK OF THIS NOTICE)
PLEASE AVOID
LEAVING LITTER
LIGHTING FIRES
DAMAGING TREES
OR PLANTS

Figure 1 A photograph of a sign found at a National Trust woodland site, UK.

Using Figure 1, explain how **one** management measure can help to protect this landscape and make it more sustainable. **(2 marks)**

....

....

....

....

> Use evidence from the photograph to help you.

3 Using example(s), describe how some people or groups are trying to conserve the biosphere at a **global scale**. **(6 marks)**

Some countries get together and develop wildlife conservation treaties to try to prevent the extinction of some wildlife and their habitats. Examples include RAMSAR in 1971, which tries to conserve wetlands, and CITES in 1973, which tries to protect endangered species. ....

....

....

....

....

....

....

....

Factors affecting biomes

HIGHER

B/A

EXAM ALERT

1 Explain how **temperature** and **precipitation** affect the location of biomes. **(4 marks)**

....

....

....

....

....

....

....

....

> Exam questions similar to this have proved tricky – be prepared! **ResultsPlus**

> Focus on temperature and precipitation. Local factors (soil, altitude, etc.) would not be relevant to the answer.

FOUNDN

E/D

2 Study Figure 1.

Figure 1 The distribution of tropical rainforests in Brazil.

Describe the **distribution** of tropical rainforests in Brazil. **(2 marks)**

....

....

....

....

HIGHER

C/B

3 Define the term 'ecosystem'. **(2 marks)**

....

....

....

....

Biosphere management tensions

FOUNDN

C

〉**Guided**〉

1 Outline **two** different management measures to conserving the biosphere. **(4 marks)**

Method 1: In Glen Affric (Scotland), the area was made into a nature reserve in the 1960s. The approach to conservation has been the reintroduction of native species (linked to the ancient Caledonian Forest), which includes the wild boar, and in future, the wolf and the beaver.

Method 2: ....

....

....

....

> Try to choose a very different method of management to set against method 1, which is local and small scale. This could be a global action (e.g. piece of legislation) or something which involves a different type of biosphere, e.g. marine or water. You could support your answer with another example if you know of a suitable one.

HIGHER

A

2 Using examples, explain some of the **challenges** and **tensions** in producing sustainable management outcomes for the biosphere. **(6 marks)**

....

....

....

....

....

....

....

....

....

....

....

The hydrological cycle

1 Study Figure 1.

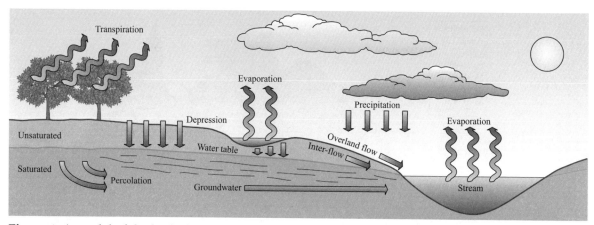

Figure 1 A model of the hydrological cycle.

> Make sure you can remember a clear meaning of all the main terms linked to the hydrological cycle as they can often appear in questions.

Which **one** of the following statements about the hydrological cycle is correct? **(1 mark)**

☐ **A** The hydrological cycle is a system of interlinked stores and transfers.

☐ **B** The hydrological cycle creates water by percolation.

☐ **C** In the hydrological cycle, water flows by transpiration.

☐ **D** In the hydrological cycle, the water table prevents the flow of water into the ground.

2 Describe **two** ways in which water moves between different stores. **(2 marks)**

1 Overland flow. This is when water flows across land, e.g. over fields, and then ends up in a stream to flow into a lake / sea store.

2 ....

....

3 Why can the hydrological cycle be described as a **system**? **(2 marks)**

....

....

....

....

Climate and water supplies

HIGHER

B

1 Study Figure 1.

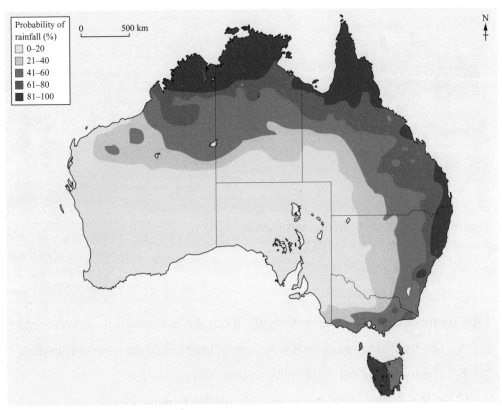

Figure 1 Rainfall reliability in Australia.

Describe the **pattern** of rainfall reliability in Australia. **(2 marks)**

....

....

....

....

> There are only 2 marks available for this question so you only need to make two points. Don't spend too long on the 'smaller' questions or you won't have enough time to complete all of the exam paper.

FOUNDN

D/C

2 Using examples, describe **two** impacts of unreliable or insufficient water supply for people. **(2 marks)**

   1 Water shortages in the UK lead to hosepipe bans which means that ....

   ....

   ....

   2 ....

   ....

Threats to the hydrological cycle

1 Outline **one** of the **impacts** of an unreliable water supply on humans.　　　**(2 marks)**

....

....

....

....

2 Using examples(s), describe some of the **threats** to a healthy hydrological cycle.　　**(4 marks)**

....

....

....

....

....

....

....

....

> Examples of threats could include industrial pollution and sewage as well as reservoir construction and deforestation. It is best to use examples of places as part of your answer, and try to give some depth of detail.

3 Using examples, explain how people may be **reducing** water supplies.　　**(6 marks)**

There are a number of serious ways in which people can negatively impact on water supplies. One of the biggest of these is reservoir and dam construction. The construction of the Three Gorges Dam and other similar projects throughout Asia, Africa and Latin America have generated many concerns (including environmental and social ones), particularly relating to local water availability. ....

....

....

....

....

....

....

....

Large-scale water management 1

FOUNDN

F/E

1 Study Figure 1.

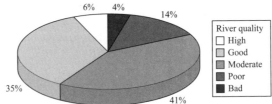

Figure 1 A chart indicating the varying quality of the UK's rivers.

Describe the water quality data shown in Figure 1. **(2 marks)**

....

....

....

....

FOUNDN

D/C

2 Using examples, explain how human activities can have an impact on **water quality**.

(4 marks)

⟩**Guided**⟩

There are many human activities that can affect the health and quality of rivers and water supplies. One of the most serious of these is intensive agriculture which is responsible for nitrate pollution and sometimes eutrophication..

....

....

....

HIGHER

A

3 Using examples, explain how industrial pollution and intensive agriculture can have an impact on **water quality**. **(6 marks)**

....

....

....

....

....

....

....

....

....

Remember that 'examples' can be of places or types of pollution / intensive agriculture.

Large-scale water management 2

FOUNDN
G/F

1 Give **one** located example of a large-scale water management scheme in the developing
world.   **(1 mark)**

....

....

FOUNDN
D/C

2 For a named large-scale project, describe some of its advantages and disadvantages.
                                                                              (4 marks)

Guided

Named project: Three Gorges Dam

Advantages: There are two significant advantages: flood control downstream (manage the
water flowing out) and also HEP (3 per cent of China's total power generation).

Disadvantages: ....

....

....

....

....

> It's always a good idea to
> include facts and figures to
> support your ideas wherever
> possible.

HIGHER
A

3 Examine the costs and benefits of a large-scale water management scheme in the developed
world.   **(6 marks)**

Guided

Early in the 1900s, the limiting factor for development of land in the western United States
was the availability of water. Damming the Colorado River, which drains snowmelt and rain
from the western side of the Rocky Mountains, was seen as a solution in the 1920s. It
provided a number of advantages ....

....

....

....

....

....

....

....

....

Small-scale water management

FOUNDN

E/D

1 What is meant by the term 'intermediate technology solution'? **(2 marks)**

....

....

....

....

HIGHER

C

2 Outline **two** ways in which intermediate technology improves water resources. **(4 marks)**

....

....

....

....

....

....

....

....

....

> Don't just give two examples here – you need to give the examples along with a brief description of how they improve water resources.

HIGHER

A

⟩ **Guided** ⟩

3 Explain how intermediate and small-scale technology can be used to help improve water resources for people in some parts of the developing world. **(6 marks)**

Improving water resources means improving the quality of water as well as making the supply of water more regular. Intermediate and small-scale technology helps to do this because it is cheap and easy to operate so local people can afford it and do not require extensive training to manage it. One example which helps irrigate crops is local rainwater harvesting ....

....

....

....

....

....

....

....

....

Coastal landforms and erosion

HIGHER
C

1   Study Figure 1.

Figure 1   An example of a coastal landform feature.

Identify the type of hard-rock landform shown in Figure 1 and briefly describe how it is formed.   **(2 marks)**

Guided

Type of coastal landform: Stump

How it is formed: ....

....

....

....

....

FOUNDN
E/D

2   Outline **one** process of coastal erosion.   **(2 marks)**

Process of coastal erosion:

....

....

....

> You need to correctly identify a type of coastal erosion **and** include developed ideas of how it works.

HIGHER
B

3   Describe how wave action can erode coastal cliffs.   **(3 marks)**

Guided

Waves strike the base of the cliffs. The hydraulic action of the water wears the rock in the cliff face away bit by bit through the process of erosion. This repeated action of waves striking the base of the cliff creates a notch ....

....

....

> Think about weathering, abrasions and attrition too.

....

Coastal landforms and deposition

HIGHER
C/B

1 Study Figure 1.

(a) Name the type of landform shown in Figure 1. **(1 mark)**

....

(b) Briefly describe how it is formed. **(3 marks)**

Figure 1 An example of a coastal landform feature.

Guided

Beaches are made up of eroded material that has been transported from elsewhere and deposited by the sea – longshore drift is often an important process ....

....

....

FOUNDN
D/C

2 Describe how a spit is formed. **(4 marks)**

....

....

....

....

....

....

....

> You could partly answer this question with a diagram as long as you don't then repeat all of the information in the text as well. Think carefully about the labels you use, as well as process arrows. Always give your diagram a title.

HIGHER
B/A

3 Explain how different types of wave can influence the profile (shape) of a beach. **(4 marks)**

....

....

....

....

....

....

....

Geology of coasts

HIGHER

B/A

Guided

1 Explain the differences between **discordant** and **concordant** coasts.   **(4 marks)**

A discordant coast is where the geology alternates between bands of hard and soft rock

whereas a concordant coast ....

....

....

....

....

....

FOUNDN

D/C

2 Explain what is meant by the term 'sub-aerial' processes?   **(2 marks)**

....

....

....

....

HIGHER

A/A*

3 For a named coastline, explain how different wave types can lead to the development of a variety of coastal landforms.   **(8 marks + 3 marks SPaG)**

Named coastline: ....

....

....

....

....

....

....

....

....

....

....

....

....

....

....

> Think about constructive and destructive waves in the context of this response, including those associated with both erosion and deposition. You will need to be selective with what features you decide to write about.

....

Factors affecting coastlines

FOUNDN

F/E

1   Study Figure 1.

> **Climate change and coastal erosion – UK**
>
> Our changing climate is causing sea levels to rise, storm patterns are becoming stronger and land levels are also changing since the last ice age. Land is beginning to fall in the south of the country (happening at a rate of about 2 mm per year).
>
> These combined processes are likely to lead to an increase in coastal erosion in some areas over the next hundred years. The natural and man-made barriers that absorb wave energy will become increasingly submerged and could be damaged as sea levels rise.
>
> *Adapted from Environment Agency website – 2012*

Figure 1 An extract from an article describing the link between climate change and coastal erosion.

Which **one** of the following statements is correct?   **(1 mark)**

☐   **A**   There are many reasons why coastal erosion might be reduced in the future.

☐   **B**   Sea level rise will flood towns and cities all over the world.

☐   **C**   The natural and man-made barriers to coastal defence are being improved.

☐   **D**   Coastal erosion will increase due to climate change and land falling in the south.

HIGHER

C/B

2   Outline **one** possible impact of climate change on coastal areas.   **(2 marks)**

....

....

....

FOUNDN

C

⟩ **Guided** ⟩

3   Using examples, explain why coasts retreat at different rates.   **(6 marks + 3 marks SPaG)**

There are many factors which affect the rate of coastal retreat. One of these is rock type – hard rocks such as those at Land's End in Cornwall erode more slowly than soft rocks, such as those near Scarborough in Yorkshire. Another factor is ....

....

....

....

....

....

....

....

....

....

> There are 3 marks available for spelling, punctuation and grammar so make sure these are really good and that you use correct geographical terminology.

Coastal management

OUNDN
F/E

1 Study Figure 1.

Figure 1 An example of coastal defences on the south coast of England.

uided

(a) Name the type of coastal defence shown in Figure 1. **(1 mark)**

Rip-rap

(b) Briefly describe how it works. **(2 marks)**

....

....

....

OUNDN
E/D

2 Outline why some places have chosen to use **hard defences** at the coast. **(2 marks)**

....

....

....

....

HIGHER
VA*

3 For a named coastal location, describe the use of sustainable and Integrated Coastal Zone Management techniques that have been used to protect the coastline.

(8 marks + 3 marks SPaG)

Named coastal location: ....

uided

Sustainable approaches may include 'do nothing' and 'strategic realignment' which are often

controversial but sometimes very effective...

....

....

....

....

....

....

....

....

....

....

Had a go ☐   Nearly there ☐   Nailed it! ☐

Rapid coastal retreat

FOUNDN
E

1   Define what is meant by the term 'coastal retreat'.   **(2 marks)**

....

....

....

....

HIGHER
B

2   Outline why 'do nothing' may be the preferred way to deal with coastal retreat in some places.   **(3 marks)**

....

....

....

....

....

....

> You can always use examples in questions like these – they will be credited.

HIGHER
A/A*

3   For a named location, explain the main problems caused by coastal retreat.
                                                          (8 marks + 3 marks SPaG)

> **Guided**

In Overstrand, Norfolk, the soft cliffs have led to rapid coastal retreat in the last few centuries. This has created a number of serious problems, including loss of houses close to the edge of the cliffs and farmland. Other problems include ....

....

....

....

....

....

....

....

....

....

....

....

River systems

FOUNDN
E/D

1 What is meant by the term 'long profile'. **(2 marks)**

....

....

....

HIGHER
B

2 Study Figure 1.

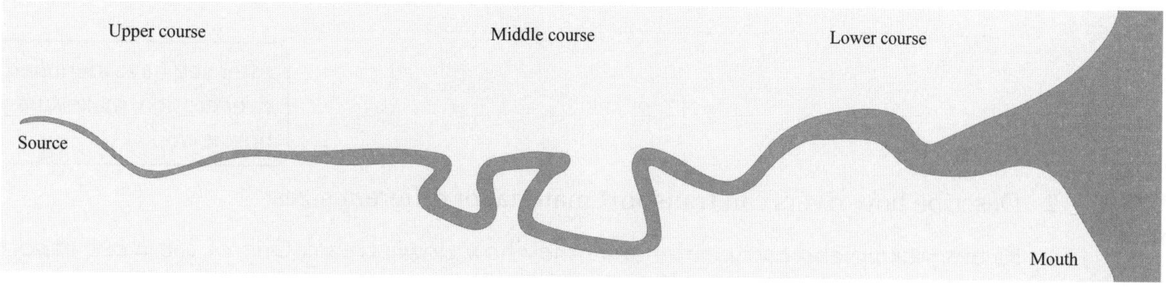

Upper course   Middle course   Lower course

Source

Mouth

Figure 1 A diagram showing the changes in a river from source to mouth.

Describe how channel **shape** and **gradient** change along the long profile of a river. **(4 marks)**

....

....

....

....

....

....

....

> Read the question carefully! This one specifically asks for 'shape' **and** 'gradient'.

FOUNDN
C

3 In a named river you have studied, explain how **width** and **depth** change from source to mouth. **(6 marks + 3 marks SPaG)**

Guided

Named River: River Horner, Somerset

Usually rivers are narrow near their source, formed of small tributaries only. As more water enters the channel it becomes much wider nearer the mouth – this is the case with the River Horner where the width increases from just a few cms to 5 metres.

....

....

....

....

....

....

Processes shaping rivers

FOUNDN
E/D
Guided

1 Describe **one** process of river erosion. **(2 marks)**

Type of river erosion:

....

....

....

....

> After you have identified a type of river erosion, make sure you explain how it works.

HIGHER
B

2 Describe how rivers can **transport** material of different sizes. **(3 marks)**

Rivers pick up and carry material as they flow downstream. One of the most important processes is traction where large boulders and rocks are rolled along the river bed. ....

....

....

....

....

HIGHER
A/A*

3 Explain the influence of **geology** and **slope processes** on river valley shape and sediment size characteristics. **(8 marks + 3 marks SPaG)**

....

....

....

....

....

....

....

....

....

....

....

....

....

....

....

> Always make sure you carefully follow the command instructions in the question. In this question you need to write about both geology **and** slope processes. Geology may include structure and rock hardness. For example, valley shape is controlled by geology – steeper valleys may be linked with harder rocks.

Upper course landforms

FOUNDN E/D

1   Study Figure 1.

Figure 1 An example of an upland river landform feature.

Guided

   (a)   Identify the type of landform shown in Figure 1.   **(1 mark)**

   Interlocking spur

   (b)   Briefly describe how it is formed.   **(2 marks)**

   ....

   ....

   ....

   ....

HIGHER A/A*

2   Using a located example, explain the erosion processes involved in the development of a **waterfall**.   **(8 marks + 3 marks SPaG)**

   ....

   ....

   ....

   ....

   ....

   ....

   ....

   ....

   ....

   ....

   ....

   ....

Remember to include all the different processes for full marks – think about all aspects of the waterfall.

Lower course landforms

FOUNDN
F/E

1   Which of the following landforms is found in the lower course of a river?   **(1 mark)**

☐ **A** Interlocking spurs

☐ **B** Waterfall

☐ **C** Plunge pool

☐ **D** Ox-bow lake

HIGHER
B

2   Describe how an ox-bow lake is formed. You may use a diagram in your answer.   **(4 marks)**

....

....

....

....

....

....

....

....

> If you have the choice of using a diagram it may be a good idea as long as you don't then repeat all of the information in the text as well. For a 'formation'-type diagram, think carefully about the annotations you might use, as well as process arrows. Always give your diagram a title.

FOUNDN
E/D

Guided

3   Briefly describe the process that forms levees.   **(2 marks)**

Levees are raised river banks made up of sediment. They are formed by ....

....

....

Causes and impact of flooding

FOUNDN
E/D

1   Outline the main factors that can **cause** rivers to flood   **(3 marks)**

....

....

....

HIGHER
C/B

2   Study Figure 2.

> ### Climate change and flood risk in the UK
>
> Some people and scientists think that changes in our climate may be leading to increased risks from river flooding, especially in towns and cities. In the last 15 years there have been a number of big floods in the UK affecting many urban areas; the results of which have caused misery to people and considerable damage to their houses. In the summer of 2007, for instance, there was very heavy rainfall that was greater than the rainfall records that were started in 1879.
>
> Because warmer air can hold more water, climate change will give the potential for stronger rainfall events. Whilst this may increase the risk of flooding in many places, flooding will, however, vary widely from location to location depending on local climatic changes that can be difficult to predict.

Figure 2 An extract from an article describing the link between climate change and flooding

Suggest how climate change may **increase** the risk of flooding in some parts of the UK.

(2 marks)

....

....

....

....

FOUNDN
C

3   Using Figure 2 and your own knowledge, explain how human activity may be worsening the impact of flooding.   **(6 marks + 3 marks SPaG)**

Guided

A warming atmosphere is allowing a greater potential for flood risk by increasing the intensity and ferocity of storm events. These lead to flashier responses, such as the floods in Devon in November 2012 when the impact was so severe the main railway line was washed away near Exeter. ....

....

....

....

....

....

....

....

> Remember that human activity will include urbanisation, changing land uses, etc. as well as possible aspects of climate change that are described in Figure 2.

Managing river floods

FOUNDN
E/D
Guided

1 (a) Name **one** type of hard (traditional) flood defence. **(1 mark)**

Flood wall

(b) Briefly describe how it works. **(2 marks)**

....

....

....

....

....

HIGHER
A/A*

2 For a named flood management scheme, examine the **costs** and **benefits** of using soft (sustainable) flood management techniques. **(8 marks + 3 marks SPaG)**

Named flood management scheme / area: ....

....

....

....

....

....

....

....

....

....

....

....

....

....

HIGHER
C/B

3 What is meant by the term 'sustainable flood defence'. **(2 marks)**

....

....

....

....

> 'Sustainable' is another term for modern or soft engineering approaches.

Threats to the ocean

1 Describe the global **distribution** of a named type of marine ecosystem.   **(2 marks)**

....

....

....

....

2 For a named type of marine ecosystem, explain the global threats it faces.
(6 marks + 3 marks SPaG)

Named ecosystem type: Coral reefs

One of the global threats faced by coral reefs is the increase in temperature of the sea due
to climate change. This leads to coral bleaching ....

....

....

....

....

....

....

....

....

....

....

....

> This is a 6 mark 'explain' question.
> You will also be assessed on your
> quality of writing as well as the quality
> of your geography. You should spend
> about 6–8 minutes thinking and
> writing this style of answer.

3 Outline **one** reason why marine ecosystems should be protected.   **(2 marks)**

....

....

....

....

Had a go ☐ Nearly there ☐ Nailed it! ☐

Ecosystem change

1 Study figure 1.

Figure 1 Rubbish in the ocean.

Describe how **one** human activity threatens marine ecosystems. **(2 marks)**

....

....

....

....

2 For a named type of marine ecosystem, outline **one** way in which it has been damaged in the past. **(2 marks)**

Named marine ecosystem: ....

....

....

....

3 For a named type of marine ecosystem, explain some of the **pressures** it is facing today. **(6 marks + 3 marks SPaG)**

⟩ **Guided** ⟩

Named marine ecosystem: Coral reefs

In many parts of the world, pollution from rivers has damaged some important coral reef ecosystems. Other pressures are linked to tourism and include divers taking souvenirs and also the noise / disruption from motorised craft in the water. Increasingly, however, the most significant pressures are those associated with climate change ....

....

....

....

....

....

....

....

Make sure you link your answer to a specific species.

Pressure on the ecosystem

FOUNDN
G/F

1   Study Figure 1.

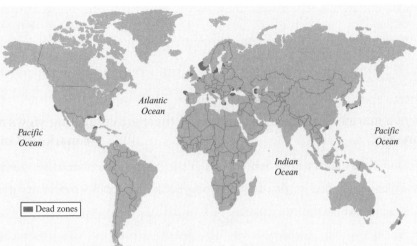

Figure 1 World map showing the location of 'marine dead zones'.

Which of the following describes the distribution of marine dead zones?   **(1 mark)**

☐   **A**   Mostly located near to the equator.

☐   **B**   Mostly located on the coastlines of developed countries.

☐   **C**   Mostly located in the middle of oceans.

☐   **D**   Mostly located on the coastlines of developing countries.

HIGHER
A/A*

2   Explain the processes that can disrupt marine food webs and nutrient cycling.   **(8 marks + 3 marks SPaG)**

....

....

....

....

....

....

....

....

....

....

....

....

....

....

....

> This is an 8 mark 'explain' question. It is important to offer a range of processes (e.g. overfishing, siltation, eutrophication) and give sufficient depth and detail, perhaps using brief examples to support your ideas.

Localised pressures

FOUNDN
E/D

1 Describe **one** example of a strategy to protect a marine ecosystem.   **(2 marks)**

....

....

....

....

HIGHER
A/A*

EXAM ALERT

2 For an example of a marine ecosystem, examine why there are conflicting views about how
it should be managed.   **(8 marks + 3 marks SPaG)**

There are conflicting views about how fish stocks in the North Sea should be managed. The
EU Common Fisheries policy has attempted to bring back fish stocks from catastrophically
low levels. ....

....

....

....

....

....

....

....

....

....

....

....

....

....

....

| Make sure your spelling, punctuation and grammar are really good and that you use accurate geographical terminology. | Exam questions similar to this have proved tricky – be prepared! **ResultsPlus** |

HIGHER
B

3 Describe what is meant by the term 'ocean health' of marine fisheries.   **(2 marks)**

....

....

....

....

Health in the context of marine resources may be composed of a number of ideas, e.g. biodiversity, damage by fishing, tourism quality, etc.

Local sustainable management

1   Study Figure 1.

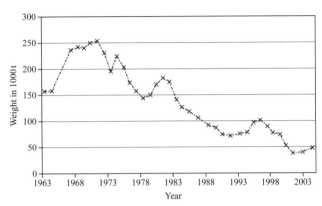

Figure 1 Changes in the weight (biomass) of North Sea cod between 1963 and 2003.

Describe the changes in the amount of North Sea cod shown on Figure 1.   **(2 marks)**

....

....

....

2   Using examples, explain how marine ecosystems can be managed in a more sustainable
way.   **(6 marks + 3 marks SPaG)**

Guided

At the core of sustainability is protecting areas for the longer term and for future

generations. In the Firth of Clyde (Scotland), for instance, a number of pressures (overfishing,

military, pollution) led to the need to produce a more sustainable future. One solution was a

no-fish zone in Lamlash Bay ....

....

....

....

....

....

....

....

3   Describe what is meant by the term 'environmental sustainability' of marine fisheries.

(2 marks)

....

....

....

....

Although this is a
knowledge recall
question, you should
aim to develop your
statement.

Global sustainable management

1   Using examples, explain how global actions are attempting to create sustainable marine ecosystems.   **(6 marks + 3 marks SPaG)**

One example of a global action is the development of Marine Protected Areas. In 2010, the world had about 6000 MPAs, which is just over 1 per cent of the world's oceans. MPAs are regions in which human activity has been placed under some restrictions in the interest of conserving the natural environment. ....

....

....

....

....

....

....

....

....

2   Study Figure 1.

Country	Number of MPAs	Total area (km²)
Denmark	24	8403
France	9	3598
Germany	6	16889
Iceland	7	79
Norway	8	80589
Spain	2	2483
UK	63	15864

Figure 1 Number and size of Marine Protected areas (MPAs) for selected countries.

Comment on the differences in MPAs (Marine Protected Areas) between the different countries shown.   **(3 marks)**

There is a wide variation in both the number and total area across the European countries in Figure 1. Spain has the smallest number of areas (2) and the UK the largest. ....

....

....

....

> Use other data from the table to look for differences in the total areas. An interesting point to make is that there is no link (or correlation) between the number of MPAs and the total areas.

Extreme climates: characteristics

1 Study Figure 1.

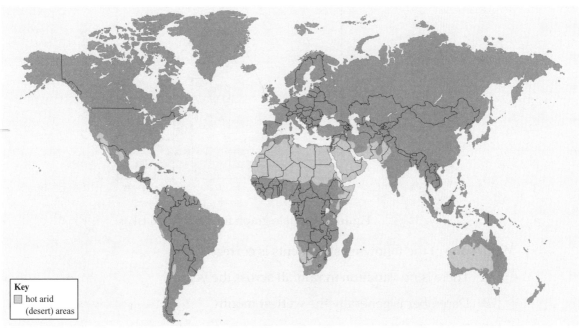

Key
☐ hot arid (desert) areas

Figure 1 The global distribution of hot arid (desert) areas.

Describe the distribution of hot arid areas. **(2 marks)**

....

....

....

....

> There are only 2 marks available for this question so you don't need to go into too much depth and detail – just mention the main points.

2 Using Figure 1, which **one** of the following statements is correct? **(1 mark)**

☐ **A** There are no hot arid areas found in the southern hemisphere.

☐ **B** The hot arid areas are mainly in North and South America.

☐ **C** Australia has little hot arid area in terms of the size of the country.

☐ **D** The biggest concentration of hot arid areas is in a belt across northern Africa.

3 Briefly outline **two** characteristics of polar regions. **(2 marks)**

1 They are dry and don't receive much precipitation.

Guided

2 ....

....

Why are extreme climates fragile?

FOUNDN
E/D

1 Study Figure 1.

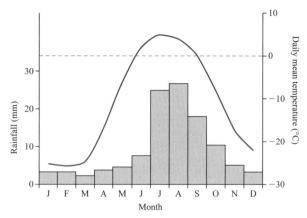

Figure 1 Climate graph for Barrow, Alaska.

Which **one** of the following statements is correct?   **(1 mark)**

☐ **A** There is no variation in rainfall across the year.

☐ **B** December is generally the wettest month.

☐ **C** Total rainfall is low throughout the year, with July and August being the wettest months.

☐ **D** The total rainfall is 1000 mm / year.

HIGHER
B

2 Describe **one** adaptation of a **plant** that is suited to an extreme climate.   **(2 marks)**

....

....

....

....

> Remember that with a 2 mark question such as this, you need to make a statement, e.g. surface layers of the soil are moist in summer (one idea) … this creates ideal conditions for sedges which are adapted to these wet and cold climates (development of idea).

HIGHER
B

3 Describe **one** adaptation of an **animal** that is suited to an extreme climate.   **(2 marks)**

The red kangaroo is adapted to the hot arid climate of Australia because ....

....

....

....

People and extreme climates

FOUNDN
F/E

1 In a polar region, give **one** way in which local buildings and farming have been developed to cope with the extreme climate. **(2 marks)**

Guided

Local buildings: In polar climates the houses have steep roofs so that snow falls off.

Farming: ....

....

....

A   B

Figure 1 Traditional clothing worn in **A** hot climates (Masai tribespeople) and **B** in cold climates (Inuit family).

HIGHER
C/B

2 Study Figure 1.

For **either** picture A **or** B, explain how people have adapted their clothing to their particular extreme climate conditions. **(2 marks)**

Picture:

....

....

....

....

> You should aim to use the picture as an idea for you answer, but you will also need to show understanding to link the picture to a particular adaptation.

FOUNDN
E/D

3 Describe **one** design feature of buildings to cope with the climate of a polar environment. **(2 marks)**

....

....

....

....

Threats to extreme climates

1 For different extreme environments, explain how life is **changing** for its people. **(8 marks + 3 marks SPaG)**

In the Sahel (hot and arid region), there has been a gradual improvement in transport and communications (especially mobile technology 'leap-frogging' traditional landlines). This has created new links for people and helped spread new ideas and even enabled local populations to find new work and migrate. Overall, there has been an introduction of more western culture and lifestyle. ....

....

....

....

....

....

....

....

....

....

....

....

'Explain', means you need to link statements to an explanation. Try to cover three ideas – one theme has been started for you in the guidance. Different extreme climates would mean polar and hot arid – so you need to cover both.

....

2 For **either** a hot arid **or** a polar region, explain the main **threats** to people and their environment. **(6 marks + 3 marks SPaG)**

Name of region: Alaska

One big threat to the environment has been oils spills, e.g. Exxon Valdez, which caused massive environmental destruction in 1989 with the loss of 750 000 barrels of oil. Immediate effects were the deaths of 250 000 sea birds and the destruction of salmon fish eggs.

....

....

....

....

....

....

....

....

....

Extreme climates: sustainable management

HIGHER

B

1   Study Figure 1.

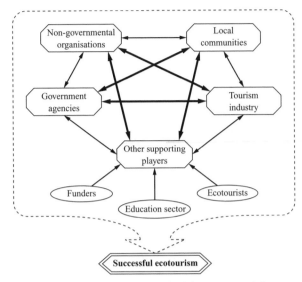

Figure 1 Players and actions needed for successful ecotourism.

In **either** a hot arid **or** a polar region, outline the importance of actions by local players in developing successful ecotourism.   **(2 marks)**

....

....

....

....

> Local actions would be local to that particular ecosystem or environment.
> Players can be groups, individuals, etc. – use the diagram to help you here.

FOUNDN

C

Guided

EXAM ALERT

2   Using examples, explain how local communities have tried to become more **sustainable** in extreme environments.   **(6 marks + 3 marks SPaG)**

In Tanzania, East Africa, people use different types of intermediate technology approaches (bottom-up) to help develop better water supplies, e.g. hand pumps, rain barrels and lined wells ....

....

....

....

....

....

....

....

> Exam questions similar to this have proved tricky – be prepared! **ResultsPlus**

Extreme climates: global management

1 Study Figure 1.

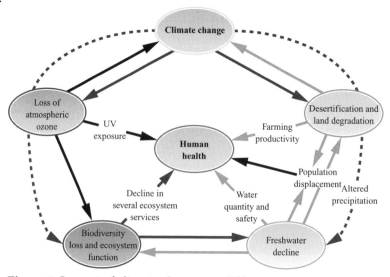

Figure 1 Impact of climate change on different systems

Using Figure 1, outline how climate change can **threaten people** in extreme climates.

(2 marks)

....

....

....

2 Explain the role of **global actions** to protect extreme environments.

(8 marks + 3 marks SPaG)

....

....

....

....

....

....

....

....

....

....

....

....

....

> Make sure your spelling, punctuation and grammar are good and that you use accurate geographical terminology.

World population growth

FOUNDN
G/F

1 Study Figure 1.

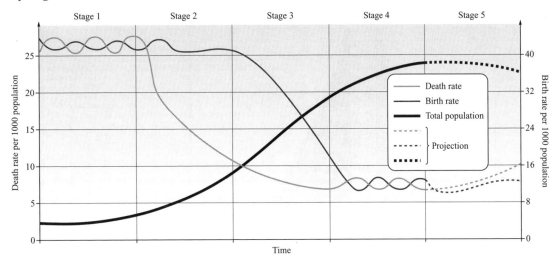

Figure 1 The demographic transition model.

In which Stage of the demographic transition model does death rate **decrease** most rapidly? **(1 mark)**

☐ **A** Stage 1

☐ **B** Stage 2

☐ **C** Stage 3

☐ **D** Stage 4

> You will have seen this model plenty of times during your course but that doesn't mean you don't need to look at it carefully if it appears in the exam as it may be slightly different from the one you've seen before.

HIGHER
C/B

2 Using Figure 1, describe what happens to **total population** in Stages 4 and 5 of the model. **(2 marks)**

Guided

In stage 4, there is a gradual increase over time ....

....

....

....

FOUNDN
E/D

3 Outline **one** reason why the rate of global population growth has slowed in recent years. **(2 marks)**

....

....

....

....

HIGHER
B

4 Describe what is meant by a population growth rate which is 'exponential'. **(2 marks)**

FOUNDN
D

....

....

....

....

Population and development

FOUNDN
F/E

1 Study Figure 1.

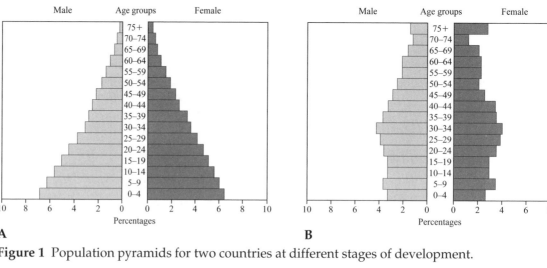

A   **B**

Figure 1 Population pyramids for two countries at different stages of development.

Which stages of the demographic transition model do the two pyramids, A and B, best represent? **(1 mark)**

☐ **A**   Stage 1 and Stage 4

☐ **B**   Stage 2 and Stage 1

☐ **C**   Stage 2 and Stage 4

☐ **D**   Stage 3 and Stage 5

FOUNDN
F/E

2 Identify the **largest** age group in Pyramid B, Figure 1. **(1 mark)**

....

....

HIGHER
B
⟩**Guided**⟩

3 Explain the **differences** in the two population pyramids shown in Figure 1. **(4 marks)**

Pyramid A has a very wide base which is typical of countries such as Nigeria and Kenya. This is most likely due to a very high birth rate and high death rate, i.e. due to poor nutrition and famine. Pyramid B, in contrast, ...

....

....

....

....

....

....

Population issues

1 Explain what is meant by the term 'dependants'. **(2 marks)**

....

....

....

....

2 Describe the shapes of the population pyramids of a country with a youthful population and of a country with an ageing population. **(4 marks)**

....

....

....

....

....

....

....

....

> For this type of question you could include a sketch diagram as part of your answer – make sure it is labelled appropriately.

3 Describe some of the problems linked to ageing populations. **(4 marks)**

An ageing population is one that has a high proportion of people of retirement age. One problem associated with an ageing population is that lots of money is needed for state pensions when there is a smaller number of working people paying taxes to cover it. Another problem is ....

....

....

....

....

....

Managing populations

FOUNDN
F/E

1   What is meant by the idea 'sustainable level of population'. **(2 marks)**

....

....

....

....

> Sustainable ideas about population numbers are complex and difficult to understand with many different viewpoints. People have their own ideas about how much space is needed, how many jobs, and available resources for a growing world population. Sustainable usually relates to a maximum idea.

HIGHER
C/B

2   Outline **one** problem caused by overpopulation. **(2 marks)**

....

....

....

....

HIGHER
A

3   Explain the reasons why some places have decided to try to manage and control their population levels. **(6 marks)**

> **Guided**

Governments try to manage population to make sure there is a balance between the resources and people available. For example, China ....

....

....

....

....

....

....

....

....

....

....

....

....

> This is a 6 mark question. It's a good idea to include details and facts wherever possible. In this instance, it would be sensible to contrast two countries which have taken different approaches.

Pro- and anti-natal policies

FOUNDN

F/E

1 Define the term 'pro-natalist policy'. **(2 marks)**

....

....

....

....

HIGHER

A

2 Examine the **advantages** and **disadvantages** of an anti-natalist policy on **one** named country. **(6 marks)**

Guided

Country: China

A reflection on China's One Child Policy since 1978 reveals several pros and cons. The main advantages include a lowering of the birth rate (the main objective of the policy) and a population structure which became more balanced as the country was lifted away from a youthful population. ....

....

....

....

....

....

....

....

....

FOUNDN

D

3 Describe **two** ways in which a country has attempted to **increase** its population. **(4 marks)**

1 ....

....

....

....

2 ....

....

....

....

> You could include two pro-natalist policies here but you could alternatively think of other ways a country may increase its population.

Migration policies

FOUNDN
G/F

1 Study Figure 1.

Figure 1 A cartoon showing one interpretation of migration into Europe.

What are the **two** main messages shown by the cartoon?   **(2 marks)**

☐ **A** Migrants are welcome within some parts of Europe.

☐ **B** Europe has an open-door policy to migration.

☐ **C** Europe has a problem with an ageing population.

☐ **D** Europe wants to prevent migration into its countries.

☐ **E** Europe has strict enforcement of restrictions to people entering its borders.

HIGHER
B

2 Describe how skills tests **or** quotas are used to manage the number of people entering a country.   **(3 marks)**

Guided

Skills tests (Life in the UK Test) are now required for permanent settlement in the UK or for British citizenship. ....

....

....

....

....

FOUNDN
D/C

3 Explain why some countries have migration policies to **encourage** migration.   **(4 marks)**

HIGHER
B

....

....

....

....

....

....

....

Types of resources

FOUNDN
G/F

1   State **one** example of a non-renewable energy source. **(1 mark)**

....

....

HIGHER
C

2   Outline **one** advantage of using renewable resources. **(2 marks)**

....

....

....

....

HIGHER
B

3   Study Figure 1.

Describe how the energy source shown in Figure 1 can create both **advantages** and **conflicts** for people. **(4 marks)**

Figure 1 Wind turbines.

Advantages to wind turbines include ....

....

....

....

Disadvantages or conflicts caused by wind turbines include ....

....

....

> Go for **at least** one advantage and disadvantage when tackling this type of question.

Resource supply and use 1

FOUNDN
E/D

1 Give **two** reasons why resource consumption is generally higher in developed countries. **(2 marks)**

Reason 1: ....

....

Reason 2: ....

....

HIGHER
B/A

2 Explain why energy supply can be a source of conflict between countries. **(4 marks)**

....

....

....

....

....

....

....

....

> Think about countries that have a plentiful supply of an energy source compared with a country that doesn't.

HIGHER
A

3 Examine what is likely to happen to global resource supply and consumption in the next 50 years. **(6 marks)**

Global consumption of energy will increase substantially in the next few decades despite efforts to reduce consumption, especially of fossil fuels. Developed countries already consume large amounts and this is likely to continue and grow. However, less developed countries ....

....

....

....

....

....

....

....

....

Resource supply and use 2

IGHER
B

1 Study Figure 1.

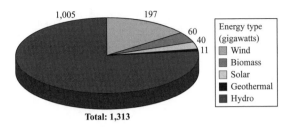

Total: 1,313

Figure 1 Global gigawatts of renewable energy by different types.

Explain why there are differences in the amounts of renewable energy used around the world. **(2 marks)**

....

....

....

....

OUNDN
D/C

2 Explain the patterns in both the **consumption** and **supply** of **one** non-renewable energy source. **(4 marks)**

> Think about the natural resources different countries have.

Guided

Non-renewable energy source: Coal

....

....

....

....

....

....

....

....

Consumption theories

FOUNDN

F/E

1   Study Figure 1.

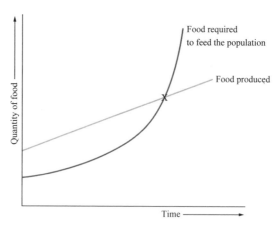

Figure 1 A model of the Mathusian population theory.

Which **one** of the following best describes what happens at point **X**?   **(1 mark)**

☐ **A** Food produced is equal to food required to feed the population.

☐ **B** Food production and food required don't cross each other.

☐ **C** Food required is less than food produced.

☐ **D** Food produced decreases over time.

HIGHER

B

2   Using Figure 1, describe how the model shows the different relationship between 'food production' and 'food required'.   **(3 marks)**

....

....

....

....

....

....

> Make sure you use ideas from Figure 1 to develop your answer. Words / phrases such as 'rising', 'falling', 'trend', 'change', 'increase' and 'over time' may also be useful.

FOUNDN

D/C

3   Describe Boserup's theory of resources and population change.   **(2 marks)**

Guided

When population increases so does food supply, thereby keeping pace with it ....

....

....

....

Managing consumption

1 What is meant by the term 'resource conservation'. **(2 marks)**

....

....

....

....

2 Outline **two** different government **policies** that are used to help manage resource consumption. **(4 marks)**

1 ....

....

....

....

2 ....

....

....

....

....

> Exam questions similar to this have proved tricky – be prepared! **ResultsPlus**

> Think about different aspects of resource consumption, such as transport and homes.

3 Explain how some governments try to manage resource consumption and try to promote sustainability. **(6 marks)**

There are several government strategies that promote recycling which is a part of sustainability. For example, in the UK many local councils have free door-side collections for glass and metal; in Sweden, 7 out of 10 people live in low-energy houses, which is due in part to the strict government planning laws and building regulations. ....

....

....

....

....

....

....

....

Potential of renewables

FOUNDN

E/D

1 Define the term 'renewable energy'. **(1 mark)**

....

....

HIGHER

B/A

EXAM ALERT

2 Study Figure 1.

Figure 1 A hydrogen bus in Iceland.

Explain how **new technologies** (such as the hydrogen bus shown in Figure 1) may resolve resource shortages. **(4 marks)**

New technologies may resolve resource shortages by developing vehicles which are run on hydrogen, therefore reducing the need for oil which will eventually run out. ....

....

....

....

....

....

....

> Exam questions similar to this have proved tricky – be prepared! **ResultsPlus**

FOUNDN

D/C

3 Describe how **new technologies** can be used to become more sustainable. **(4 marks)**

....

....

....

....

....

....

....

....

> The hydrogen economy might be useful here as a strategy in terms of energy, as could more energy conservation methods which are based around better technology.

Changing employment patterns

FOUNDN
E/D

1 Study Figure 1.

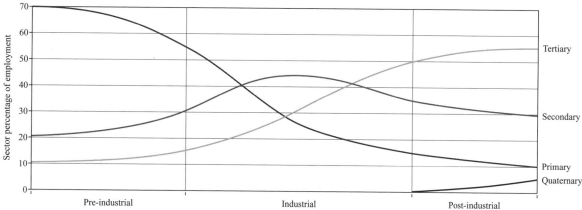

Figure 1 Clark Fisher model.

What happens to **primary** and **quaternary** employment during the post-industrial period?   **(2 marks)**

Guided

Primary: *There is an overall decrease in the number of people employed in this sector, from around 15 per cent to 10 per cent.*

Quaternary:

....

....

FOUNDN
C

HIGHER
B

2 Study Figure 1. Outline **one** reason why the proportion of people in secondary employment falls in the post-industrial period.   **(2 marks)**

....

....

....

....

HIGHER
A

3 Explain how the Clark Fisher model can be used to explain changes in **employment structure** in countries at different levels of development.   **(6 marks)**

....

....

....

....

....

....

....

....

....

....

This is a 6 mark question. Different countries should be chosen at different levels of development, i.e. high, medium and low. Try to support your ideas with data and statistics whenever possible.

Had a go ☐   Nearly there ☐   Nailed it!

Employment sectors

1 Study Figure 1.

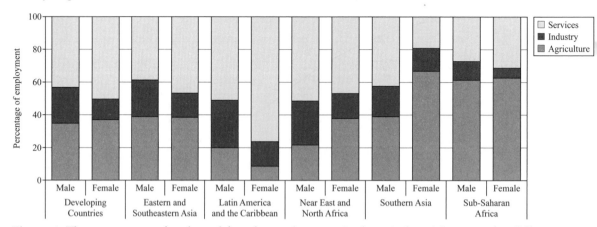

Figure 1 The percentage of male and female employment in three industrial sectors for different regions of the world.

Describe the differences in employment shown in Figure 1 between **Latin America / Caribbean** and **Sub-Saharan Africa**.   (3 marks)

....

....

....

....

....

....

> You need to look at the two different regions **and** at the differences between men and women. If possible use figures to make comparisons. Note this is a 'describe' and not an 'explain' question so you don't need to give reasons.

2 Describe the changes in **employment** that take place as countries develop.   (4 marks)

As a country develops, a wider variety of employment becomes available to people and the

percentage of people employed in different sectors of employment changes. According to

the Clark Fisher Model ....

....

....

....

....

....

....

Impact of globalisation

HIGHER
B

1 Choose **one** organisation and explain how it may be involved in the process of globalisation. **(3 marks)**

Chosen organisation / TNC: ....

....

....

....

....

....

> The idea of process is key in this question. Organisations may lend or invest money or help with new projects in different countries.

FOUNDN
D/C

2 Describe how the development of new industries in **developing** countries can bring **advantages** and **disadvantages**. **(4 marks)**

....

....

....

....

....

....

....

....

....

HIGHER
B

3 Describe some of the **positive impacts** that globalisation has had on some groups of people. **(4 marks)**

Guided

One positive impact has been an increase in the number of women in paid employment.

....

....

....

....

....

....

....

International trade and capital flows

1 Study Figure 1.

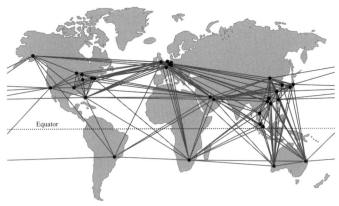

Figure 1 A computer generated map showing global airliner routes.

Which one of the following best describes the main patterns of global airliner routes in Figure 1? **(1 mark)**

☐ **A** Most are in the northern hemisphere and run in an east-west / west-east direction.

☐ **B** Most are in the southern hemisphere and run in an east-west / west-east direction.

☐ **C** Most run in a north-south / west-east direction between the northern and southern hemisphere.

☐ **D** Routes are equally distributed across the world.

FOUNDN

D

2 Describe **one** of the processes that is linked to globalisation. **(2 marks)**

....

....

....

> The focus here is on 'processes' – make
> sure you read all questions carefully.

HIGHER

A

3 Explain some of the factors that have led to the expansion of international trade in the last 50 years. **(6 marks)**

⟩ **Guided** ⟩

Globalisation has been strongly linked to international trade and one of the main backbones for this has been the advancement in various forms of technology 'shrinking space'. ....

....

....

....

....

....

....

....

....

TNCs: secondary sector

OUNDN

G/E

1 State **two** characteristics of a TNC (Transnational Corporation).   **(2 marks)**

   1 ....

     ....

   2 ....

     ....

IGHER

A

Guided

EXAM ALERT

2 Using an example of **one** named TNC in the secondary sector, explain the distribution of its global operations.   **(6 marks)**

   Named TNC: Toyota

   Toyota is the largest car manufacturer in the world. In 2013, it produced 8.7 million vehicles and globally employs 330 000 workers. It has operations around the world ....

   ....

   ....

   ....

   ....

   ....

   ....

   ....

   ....

   ....

   ....

   ....

> Exam questions similar to this have proved tricky – be prepared! **ResultsPlus**

IGHER

B

3 Outline some of the processes associated with the idea of 'global shift'.   **(2 marks)**

   ....

   ....

   ....

   ....

> When you're revising, it's very useful to produce a list of key terms (such as 'global shift') and their definitions as they will crop up in many questions.

TNCs: tertiary sector

FOUNDN

G/F

1 Give an example of a TNC (transnational corporation) that operates in the tertiary sector. **(1 mark)**

....

....

FOUNDN

D/C

2 Describe where a TNC you have studied in the **tertiary sector** operates. **(4 marks)**

....

....

....

....

....

....

....

....

> This is asking about which countries / cities a TNC operates in and what is done where. For example, where is the headquarters? Where does it source or manufacture its products?

HIGHER

B

⟩ **Guided** ⟩

3 Explain why some TNCs in the tertiary sector have moved operations overseas. **(4 marks)**

Movement of administrative functions, especially customer services / call centres has been popular for many TNCs who provide financial services, e.g. HSBC. There are several reasons for this type of move overseas ....

....

....

....

....

....

....

What is development?

HIGHER
C/B

1   Study Figure 1.

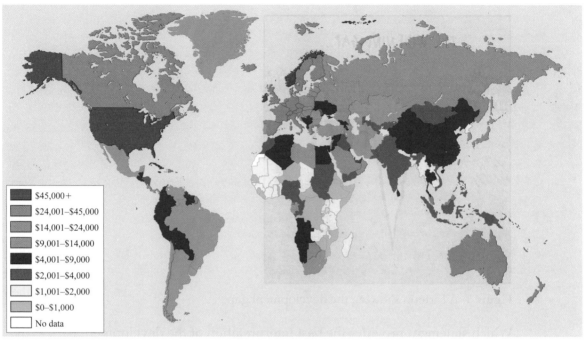

Figure 1 A global map of Gross Domestic Product (GDP) per person, 2008.

Describe the global pattern of GDP shown in Figure 1.   **(2 marks)**

....

....

....

....

> 'Pattern' and 'distribution' questions can often be linked to world
> maps. It's really important that you are familiar with some basic world
> geography of places and countries so that you can make descriptions.

FOUNDN
G/E

2   Study Figure 1. Which one of these regions has the **lowest** level of GPD per
person?   **(1 mark)**

☑ **A**   Africa

☐ **B**   South America

☐ **C**   Europe

☐ **D**   Asia

FOUNDN
D

3   Explain what is meant by the term 'GDP per capita'.   **(2 marks)**

GDP per capita means Gross *Gross Domestic Product per*
person - the total ~~both~~ wealth of a
country ① per person

....

The development gap

1   Study Figure 1.

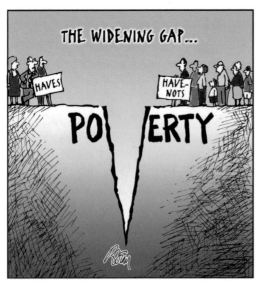

Figure 1 A cartoon showing the development gap.

> Cartoons can be quite tricky to understand so make sure you look at them carefully.

Which statement provides the best interpretation of the development gap?   **(1 mark)**

☐ **A**   All parts of the world are equal.

☐ **B**   People can fall down into poverty.

☑ **C**   There is a big gap between the haves and the have nots.

☐ **D**   The rich world is developed and is also taking resources from the poor world.

2   Explain how the HDI (Human Development Index) can be used to help understand levels of development.   **(3 marks)**

The HDI has a number of different variables, for example, life expectancy, education and income. Countries are ranked from 0 to 1. 1 being the highest and 0 being the lowest

3   Describe how the development gap has changed **over time**.   **(4 marks)**

The development gap is an expression used to describe the gulf between the rich and poor world, or the countries that are most industrialised and those which remain locked in poverty. Generally the development gap has got wider ....

....

....

....

....

....

....

Development

**HIGHER
C/B**

1 Study Figure 1.

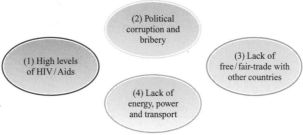

Figure 1 Some of the key development problems in parts of Sub-Saharan Africa.

Choose **one** of the development problems in Figure 1, and briefly outline why it is a barrier to development. **(3 marks)**

Guided

Development problem: (2) Corruption

TNCs and foreign investors will be reluctant to invest in countries with corrupt governments because ....

....

....

....

**FOUNDN
D**

2 For **one** country in Sub-Saharan Africa you have studied, describe some of its development problems. **(4 marks)**

....

....

....

....

**HIGHER
A**

3 For **one** country in Sub-Saharan Africa you have studied, explain the barriers to development. **(6 marks)**

....

....

....

....

....

....

....

....

....

....

....

> This is a 6 mark 'explain' question so you need to make links between comments. For example, high numbers of people with HIV / Aids leads to a less healthy workforce and lower overall productivity.

Theories of development

1 Study Figure 1.

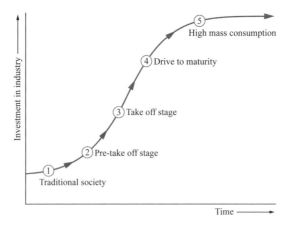

Figure 1 Rostow's modernisation and development theory

Comment on how Rostow's theory can be used to help understand economic growth over time. **(4 marks)**

....

....

....

....

....

....

....

....

> Use the diagram (Figure 1) to help you answer questions like these, but try not to just lift information with no comment or explanation.

2 Describe some of the **problems** with development theories. **(4 marks)**

....

....

....

....

....

....

....

....

Regional disparity

1 Study Figure 1.

Province (in China)	Population (millions)	GDP for the province (US$ billion)	Economic growth rate (%)
Guangdong	105	838	10.0
Henan	94	427	11.6
Chongqing	28	145	16.5
Guizhou	35	89	15.0
Tibet	3	10	12.6

Figure 1 Regional differences of GDP for selected provinces within China (2011).

Describe one difference in variations of **GDP** and one difference in **economic growth rates** shown in Figure 1. **(2 marks)**

....

....

....

2 Using Figure 1, comment on the regional differences in economic development. **(3 marks)**

....

....

....

....

....

....

> Exam questions similar to this have proved tricky – be prepared! ResultsPlus

> Use data in the table to support your ideas. 'Comment on' invites you to look at the data and then use some of your own ideas to make sense of the development geography.

3 Explain why levels of development can vary within **one** country. **(4 marks)**

In China there are a number of important factors which include historical significance, physical geography and remoteness, amount of infrastructure and locations of main cities, etc. Many of these factors are in fact linked to each other...

....

....

....

....

....

Types of development

FOUNDN
E/D

1 What is meant by 'top-down development project'? **(2 marks)**

....

....

....

....

> There is no need to give an example of a named top-down project here, unless it helps reveal something about scale which develops your answer.

HIGHER
A

2 Using a named example, explain the impact of **one** top-down project on different **groups** of people. **(6 marks)**

....

....

....

....

....

....

....

....

....

....

....

....

FOUNDN
D/C

3 Describe the main **differences** between top-down and bottom-up development projects. **(4 marks)**

Guided

One of the most important differences is that of geographical scale. Top-down tend to be very large scale, e.g. a large dam covering 1000s of km², whereas bottom-up are generally much smaller and more locally based. ....

....

....

....

....

Industrial change in the UK

FOUNDN
G/F

1   Study Figure 1.

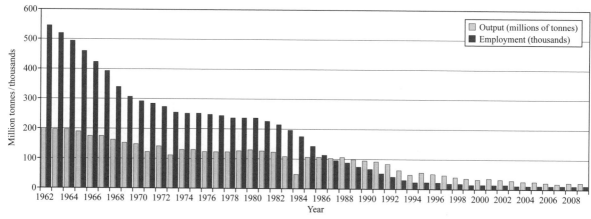

Figure 1 Changes in output and employment in the UK coal industry, 1962–2010.

Which statement best describes the changes in both output and employment?   **(1 mark)**

☐   **A**   Both declined over time.

☐   **B**   Gradual increase in coal output and employment between 1962 and 2010.

☐   **C**   Rapid decline in employment was matched by a similar drop in coal output.

☐   **D**   Between 1962 and 1992, output slowly decreased and then dropped even more, up to 2010.

> Students often rush through multiple choice options – make sure you read each one carefully to be sure you select the right option.

HIGHER
B

Guided

2   Suggest reasons for the **decline** in the number of coal miners in the UK.   **(3 marks)**

Globalisation has been an important driver to the changes in much employment in this primary extraction industry, especially the availability of cheaper sources overseas, e.g. eastern Europe where labour is relatively cheap. ....

....

....

....

....

UK employment

1 Study Figure 1.

	1973	1979	1981	1990	2010
Agriculture and fishing	1.9	1.6	1.6	1.4	1.5
Manufacturing	34.7	31.3	28.4	20.5	8.2
Distribution, catering and hotels	17.4	18.4	19.1	21.4	21.3
Banking and finance	6.4	7.2	7.9	15.2	20.3

Figure 1 Changes in employment structure of the UK (%), selected types, 1973–2010.

Between which years did **manufacturing** see the biggest decline? **(1 mark)**

☐ **A** 1973–1979   ☐ **C** 1981–1990

☐ **B** 1979–1981   ☐ **D** 1990–2010

2 Comment on the shift towards more part-time and temporary jobs in the UK in recent decades. **(2 marks)**

....

....

....

....

> Changes are to do with employers wanting a more flexible labour force (it may be easier to offer short-term contracts for instance), plus the impact of the recession. Also, women who return to work may want a greater opportunity to mix childcare and work, so part-time is a good option for them.

3 Outline the importance of changes in the way in which people work in the UK, such as home-working and self-employment. **(6 marks + 3 marks SPaG)**

⟩ **Guided** ⟩

Flexible working practices are becoming increasingly important for the employee and the business. With changes in technology, for instance, meetings can be conducted online (e.g. using Skype) and also training (e.g. Adobe Connect). This allows people to work from home or even on the train. ....

....

....

....

....

....

....

....

UK regions and employment

FOUNDN
E/D

1 Explain why **one** industry has chosen to locate in **one** region of the UK. **(2 marks)**

Region: ....

Industry: ....

....

....

HIGHER
C/B

2 Describe the main changes in employment in **one** UK region over the last 40 years. **(3 marks)**

Region: ....

Guided

In the 1970s, a large number of people were employed in .... but this

declined ....

....

....

....

HIGHER
A/A*

3 For **two** contrasting areas of the UK, explain the differences in their industries and
workforce. **(8 marks + 3 marks SPaG)**

Named areas: ....

....

....

....

....

....

....

....

....

....

....

....

....

> Remember, there are three additional
> marks allocated to 8 mark questions
> for spelling, punctuation and
> grammar – be careful and try to use as
> many geographical terms as you can!

Environmental impact of changing employment

1 Study Figure 1.

(1) Land left derelict with empty buildings

(2) Pollution remains in land and watercourses

(3) Goods manufactured overseas are now transported to UK

Figure 1 Some of the negative impacts of deindustrialisation.

Choose **one** of the impacts in Figure 1 and describe how it creates a **negative** impact on places **or** the environment. **(2 marks)**

Impact: ....

....

....

2 Outline the **positive** environmental impacts of deindustrialisation for a named urban area in the UK. **(3 marks)**

Named urban area: ....

> **Guided**

The loss of traditional industry in .... has meant initiatives to improve the

environment in this area. ....

....

....

....

....

3 Describe the environmental impacts of deindustrialisation for an urban area of the UK you have studied. **(6 marks + 3 marks SPaG)**

Named urban area: ....

....

....

....

....

....

....

....

....

....

....

....

....

Environmental impacts can be both positive and negative. Try and provide a reasonable balance in your answer.

Greenfield and brownfield development

FOUNDN
F/E

Guided

1 Describe **one** feature of a greenfield site. **(2 marks)**

Greenfield sites tend to be areas of natural vegetation ....

....

....

....

HIGHER
B

2 Describe **one** economic diversification of a UK urban area. **(2 marks)**

....

....

....

....

FOUNDN
C

3 Describe the **costs** and **benefits** of regeneration of a brownfield site within the UK that you have studied. **(6 marks + 3 marks SPaG)**

....

....

....

....

....

....

....

....

....

....

....

....

....

> Remember to check that your spelling, punctuation and grammar are really good.

> Brownfield sites are those which have previously had industry or housing on them. 'Costs' and 'benefits' is another way of saying advantages and disadvantages.

New employment areas

1 Study Figure 1.

Green sector	UK market value (£ million)	Employed in 2008	Projected employed by 2014
Environmental	22	192 000	235 000
Renewable energy	31	260 000	430 000
Low carbon technologies	53	440 000	670 000

Source BERR (2009)

Figure 1 Data on the UK green economy.

Using Figure 1, outline the importance of green sector work to the UK economy.   **(2 marks)**

> **Guided**

The total number of people employed in the green sector is ....

....

....

2 State **two** areas of UK employment which have seen a rise in foreign workers over the last few years.   **(2 marks)**

1 ....

2 ....

3 Explain the increasing importance of the 'green' employment sector to the UK economy.   **(8 marks + 3 marks SPaG)**

....

....

....

....

....

....

....

....

....

....

....

....

....

....

....

....

> Green sector employment is set to be a significant part of the UK economy, especially in the context of 2020 emissions targets and increasing fuel costs. Industries and householders are keen to cut energy costs for instance, and this provides a range of new employment opportunities.

Urban change in the UK

FOUNDN
F/E

1   Outline what is meant by the idea of **population decline** in an area.   **(2 marks)**

....

....

....

....

HIGHER
B

2   Study Figure 1.

London's population up by 12% in 10 years
The Office for National Statistics has revealed that the capital contained 8 173 900 inhabitants (March 2011). This represents a rise of 12% since the last census was conducted in 2001, the largest of any region of England and Wales. The fastest-growing boroughs were Tower Hamlets (up 26% to 254 100), Newham (up 23% to 308 000) and Hackney (up 19% to 246 300).

Figure 1 A recent online blog extract.

Suggest reasons why there has been a significant **increase** in the population of London.

   (3 marks)

....

....

....

....

....

....

> There may be both demographic and social factors that are important.

HIGHER
B

3   Describe **one** process that has caused **either** economic **or** population decline to a named urban area of the UK.   **(2 marks)**

Guided

Named urban area: Liverpool

Liverpool is one of the largest cities in the United Kingdom. Its economy is dominated by

service sector industries and the recent government cuts have meant ....

....

....

....

....

Changes in urban areas

FOUNDN
D

1 Study Figure 1.

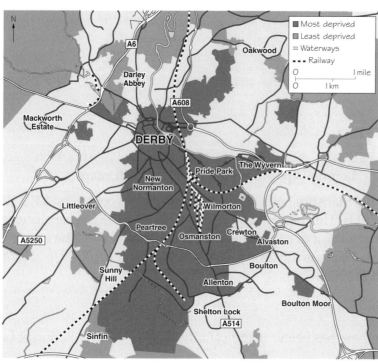

Figure 1 A GIS map of Derby showing an Index of Multiple Deprivation (IMD), 2011.

Which of these statements best describes the pattern of deprivation shown?   **(1 mark)**

☐ **A** There is no deprivation.

☐ **B** Most of the deprivation is in the north and west.

☐ **C** The highest levels of deprivation are found in a central-south corridor.

☐ **D** The highest levels of deprivation are in the peripheral areas of Derby.

HIGHER
B

2 Using Figure 1, describe the main pattern of multiple deprivation in Derby in 2011.   **(2 marks)**

....

....

....

....

> Use compass directions to help with your description, e.g. in the north there is ... etc.

FOUNDN
D

Guided

EXAM ALERT

3 What is meant by the term **multiple deprivation**.   **(2 marks)**

Multiple deprivation means that an area has a shortage of things necessary for people to lead healthy lives such as ....

....

....

> Exam questions similar to this have proved tricky – be prepared! **ResultsPlus**

Rural settlements

**HIGHER
B**

1 Study Figure 1.

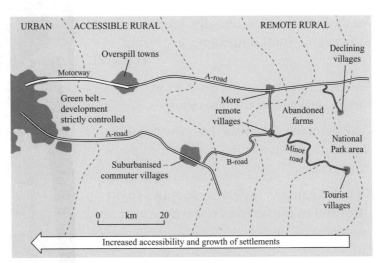

Figure 1 A diagram showing the range of rural areas that is typically found in the UK.

Explain how some of the rural settlements shown in Figure 1 may have developed. **(3 marks)**

Guided *Accessibility may be a key factor in the development of rural settlements, especially being close to main roads that offer links to places of work in towns and cities. Other factors ....*

....

....

**HIGHER
C/B**

2 Outline **one** reason why accessible rural areas of the UK have increasing populations.

(2 marks)

....

....

....

**FOUNDN
C**

3 Using examples, explain why some coastal towns have a large number of retired people living there. **(6 marks + 3 marks SPaG)**

....

....

....

....

....

....

....

....

....

> Remember to include a variety of reasons
> here – why do older people move to
> these areas and why do younger people
> move away or choose not to move there?

....

Contrasting rural areas

FOUNDN
F/E

1   Study Figure 1.

Which of these statements gives the best reason why many remote rural settlements are losing services? **(1 mark)**

☐ **A**   People prefer to use the services in the large urban settlements that are close.

☐ **B**   The permanent population is declining so fewer people live in these areas to support the services.

☐ **C**   Tourists bring in their own supplies so do not need to use services when they visit remote settlements.

☐ **D**   The services in remote rural settlements are generally of a poor quality so people choose not to use them.

> Remember to read all the options in multiple choice questions very carefully – it's easy to trip up and just select the first one you read that sounds right.

Figure 1 Many shops in remote rural settlements have closed down in recent years.

HIGHER
A*

2   Explain the variations in **quality of life** for **two** rural areas of the UK.

(8 marks + 3 marks SPaG)

Guided

An area of accessible countryside is .... whereas an area of remote

countryside is ....

....

....

....

....

....

....

....

....

....

....

....

....

....

....

> Make sure your spelling, punctuation and grammar are really good and that you use accurate geographical terminology.

Impact of housing demand

**FOUNDN
E/D**

Guided

1 Describe **one** environmental impact of rising demand for housing in urban areas.   **(2 marks)**

In some urban areas, e.g. Bristol, developers are now buying greenfield locations in the form

of people's back gardens. This has a negative impact on the environment because ....

....

....

**FOUNDN
C**

2 Using examples, explain how some **urban** areas have become improved through
regeneration and rebranding.   **(6 marks + 3 marks SPaG)**

....

....

....

....

....

....

....

....

....

**HIGHER
A***

3 Using examples, examine the success of different **urban** rebranding strategies.
   (8 marks + 3 marks SPaG)

....

....

....

....

....

....

....

....

....

....

....

....

....

> Note the focus here is on 'success' – you
> need to say whether or not the strategies
> have worked and again support your ideas
> with evidence. It would probably be a good
> idea to write about two or three different
> strategies, ideally from different places.

Making rural areas sustainable

FOUNDN
F/E

1   Describe **one** planning policy that tries to conserve valuable landscapes.   **(2 marks)**

....

....

HIGHER
A*

2   Examine the success of planning policies in both **conserving landscapes** and allowing rural
    economic development.   **(8 marks + 3 marks SPaG)**

....

....

....

....

....

....

....

....

....

....

....

....

> Development and conservation issues are difficult to
> manage so it is often the job of government and local
> planners to promote certain developments, e.g. small
> business parks and rural broadband, whilst preventing
> the building of too many second homes etc.

FOUNDN
C

3   Using examples, explain how some **rural** areas have become improved through
    development schemes or projects.   **(6 marks + 3 marks SPaG)**

Guided

In some rural areas, e.g. parts of Cornwall, the government has tried to promote development
through the provision of high-speed broadband. 'Superfast Cornwall' aims to deliver superfast
fibre broadband to 80 per cent of businesses in Cornwall and the Isles of Scilly by 2014.

....

....

....

....

....

....

....

....

Global trends in urbanisation

1 Study Figure 1.

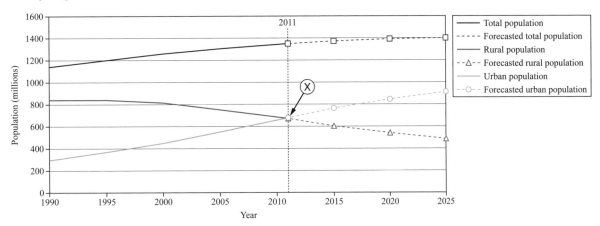

Figure 1 Urbanisation trends in China, 1990–2025.

What happens to China's population at point **X**? **(1 mark)**

☐ **A** The rural population exceeds the urban population.

☐ **B** The urban population and the rural population are equal.

☐ **C** The forecasted urban population begins to decline.

☐ **D** The population levels remain constant.

2 Study Figure 1. Describe what happens to **total population** between 1990 and 2025. **(2 marks)**

....

....

....

....

> Use data from the graph when you are describing the trend.

3 Study Figure 1. Suggest **one** reason for the trend in forecasted **urban** population (2012–2025). **(2 marks)**

....

....

....

....

4 What is meant by '**urbanisation**'? **(2 marks)**

Guided

An increasing proportion of a country's population ....

....

....

....

Megacities

FOUNDN

F/E

1 Which statement best describes the population characteristics of a 'megacity'? **(1 mark)**

☐ **A** A city where the population is more than 5 million.

☐ **B** A city with a population of 8–10 million.

☐ **C** A city with a population of more than 10 million.

☐ **D** A city with a population of more than 20 million.

HIGHER

C/B

2 Study Figure 1.

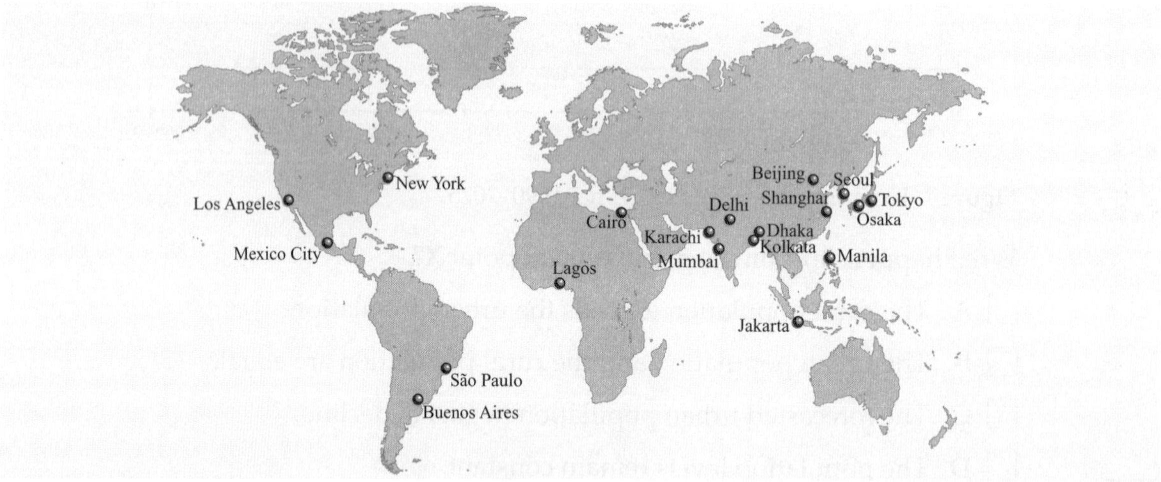

Figure 1 The distribution of megacities in 2008.

Comment on the distribution of megacities (2008) shown in Figure 1. **(3 marks)**

....

....

....

....

....

> A 'comment on' question requires you to use ideas from the resource, so don't just say where megacities are but give a reason for why you think they are found where they are.

FOUNDN

C

3 Explain the processes that have led to the rapid growth of megacities in some parts of the world. **(6 marks + 3 marks SPaG)**

⟩ **Guided** ⟩ A key growth driver in South-East and East Asia is increasing rural-urban migration which means cities such as Shanghai in China and Manila in the Philippines have grown rapidly.

....

....

....

....

....

....

....

....

Urban challenges: developed world

OUNDN
F/E

1 Study Figure 1.

Describe **one** environmental problem that this urban area may face. **(2 marks)**

....

....

....

....

Figure 1 A large urban area in Taiwan, South-East Asia.

IGHER
B

2 Use **evidence** from Figure 1 to provide reasons for the growth of this urban area in recent years. **(3 marks)**

....

....

....

....

....

....

> There are a number of clues in the photograph including the port / river / estuary facilities and the development of high-rise buildings to cater for a growing population.

OUNDN
E/D

3 Study Figure 2.

Explain how cities in the developed world are trying to solve one of the problems they face. **(3 marks)**

uided

Problem: Traffic pollution and congestion

Possible solution: ....

....

....

....

....

....

Figure 2 Congestion charging in London.

Urban challenges: developing world

FOUNDN
F/E

1 What is meant by the term 'informal economy'. **(2 marks)**

....

....

....

HIGHER
C/B

2 Describe **one** way that planners are trying to reduce pollution in rapidly expanding cities.
(2 marks)

....

....

....

> You could use either developing world cities (such as Mexico City) or developed world cities (such as Masdar City) in answering this question as it doesn't specify which.

HIGHER
A*

3 With reference to named examples, compare the range of challenges facing urban areas in the developing and developed world. **(8 marks + 3 marks SPaG)**

Guided

Developing world cities such as .... are home to many shanty settlements. Because of extreme overcrowding and poor-quality housing, there are often unsanitary conditions, a lack of clean water or sewage disposal which means there is a lot of disease.

....

....

....

....

....

....

....

....

....

....

....

....

> Make sure your spelling, punctuation and grammar are good and that you use accurate geographical terminology.

Reducing eco-footprints

HIGHER
C/B

1 Describe what is meant by the idea of an **eco-footprint**. **(2 marks)**

....

....

....

HIGHER
B

2 Study Figure 1.

UK city	Eco-footprint ('Planets consumed')
Plymouth	2.78
Wolverhampton	2.84
Liverpool	2.92
Newcastle upon Tyne	3.01
London	3.05
Southampton	3.27
St Albans	3.51

Lower eco-footprint
BETTER THAN AVERAGE

Higher eco-footprint
WORSE THAN AVERAGE

Figure 1 Eco-footprint rating table selected UK cities, average = 3.01.

Comment on why eco-footprints can vary from place to place. **(3 marks)**

....

....

....

....

....

> Eco-footprints are made up of a number of lifestyle factors, e.g. energy use, transport, etc. This may help you to understand why there could be variations.

FOUNDN
C

3 For a named city in the developed world, explain how it is trying to **reduce** its energy and waste. **(6 marks + 3 marks SPaG)**

Named city: ....

Guided

It tries to reduce energy through things such as ....

....

....

....

The city also tries to reduce waste by ....

....

....

....

....

....

Strategies in the developing world

FOUNDN

C

1 Using examples, explain some of the **social** and **environmental** problems facing cities in the developing world. **(6 marks + 3 marks SPaG)**

Guided

Many of the social and environmental problems in the developing world have their roots linked to uncontrolled urbanisation and overcrowding with limited facilities and infrastructure. Kibera outside Nairobi in Kenya is thought to be home to 1 million (illegal) settlers. It's an unplanned shanty settlement where there are several social problems – no schools, no pumped water, no power and no hospitals. ....

....

....

....

....

....

....

HIGHER

A*

2 Using examples from the developing world, explain how some cities are trying to improve the quality of life for their residents. **(8 marks + 3 marks SPaG)**

....

....

....

....

....

....

....

....

....

....

....

....

....

....

....

....

> Improvements from urban planning and self-help schemes would be ideas that you might write about. Remember that quality of life may relate to a number of different factors including life expectancy, sanitation, water supply, housing, transport etc. This means that there are a range of examples that can be used.

Rural economies

IGHER
C/B

1 Study Figure 1.

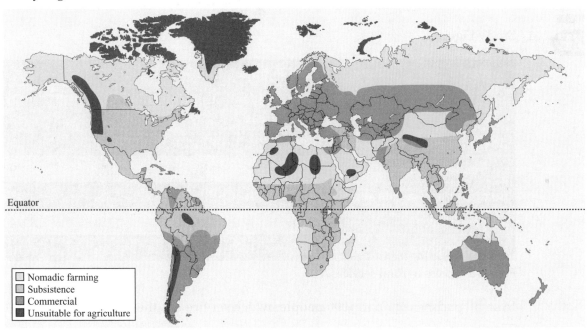

Figure 1 Main types of farming around the world.

Describe the world distribution of **subsistence** farming. **(3 marks)**

Guided

Most subsistence farming is in the southern hemisphere or just north of the equator. The

continents with a high proportion of subsistence farming are ....

....

....

....

....

OUNDN
G/F

2 Which of the following is the best description of the difference between commercial and subsistence farming? **(1 mark)**

☐ **A** Both produce similar goods, but commercial farms are much larger.

☐ **B** Commercial farms produce many types of output, subsistence farms only produce crops.

☐ **C** Commercial farms are highly mechanised, subsistence farms are not mechanised.

☐ **D** Commercial farms make a profit for the owners, subsistence farms primarily feed the owners.

OUNDN
E/D

3 Outline **one** reason why subsistence farming is still the main type of agriculture in many developing countries. **(2 marks)**

....

....

....

....

> Remember to focus on outlining just one reason here – in other words, go into detail on the one reason and do not just list reasons.

Rural challenges: developed world

FOUNDN
F/E

1 Study Figure 1.

Service	Parish population			
	Average for rural areas	Up to 1000	1000–3000	Over 3000
no evening bus service	39%	69%	41%	23%
no bus service at any time	9%	13%	8%	9%
no taxi or dial-a-ride service	53%	71%	57%	43%
no post office	16%	48%	11%	4%
no village shop	19%	49%	12%	10%

Figure 1 Access to rural services.

In small parishes of up to **1000 people**, which service has the worst score? **(1 mark)**

☐ **A** no bus service

☐ **B** no post office

☐ **C** no village shop

☐ **D** no taxi / dial-a-ride service

HIGHER
B

2 Study Figure 1. Describe the **access to services** for different sizes of parish. **(2 marks)**

....

....

....

....

> Include figures from the table as part of your answer to prove the points being made.

HIGHER
B

Guided

3 Outline how **tourism** can bring pressures to a rural area in the **developed** world. **(2 marks)**

There are obvious pressures such as overcrowding and traffic problems on rural roads especially during busy periods, such as weekends and school holidays. Also ....

....

....

Rural challenges: developing world

HIGHER

A*

EXAM ALERT

1 In a named rural area in the developing world, examine some of the challenges that exist.   **(8 marks + 3 marks SPaG)**

Named rural area:..

....

....

....

....

....

....

....

....

....

....

| Exam questions similar to this have proved tricky – be prepared! **ResultsPlus** |

....

....

| There are a number of challenges that you could write about here: rapid and uncontrolled rural-urban migration, isolation, lack of infrastructure, natural hazards, etc. |

....

....

....

FOUNDN

C

Guided

2 Using examples, explain some of the problems for rural areas in developing countries.   **(6 marks + 3 marks SPaG)**

In some parts of the developing world, such as in the countries around the edges of the Sahara Desert in Africa, desertification is increasing so more and more land is becoming infertile which means it can't be used to grow crops or to graze animals on. Another problem in ....

....

....

....

....

....

....

....

| You should have studied a named rural area in a developing country. Using these facts and figures to support your points is important in these questions. |

Rural development projects

1 Using examples, examine initiatives that have been used to improve quality of life in rural areas of the **developing** world. **(8 marks + 3 marks SPaG)**

There are a range of different initiatives that have been successfully used to improve people's quality of life in different parts of the developing world. These include micro-finance projects, mobile health services and various local education initiatives. ....

....

....

....

....

....

....

....

....

....

....

....

> Make sure that your spelling, punctuation and grammar are really good.

> For these questions you will need to develop range, depth and detail in your answer.

2 Using examples, explain the roles of **different groups** in development projects in rural areas. **(6 marks + 3 marks SPaG)**

....

....

....

....

....

....

....

....

....

....

....

> Make sure that your spelling, punctuation and grammar are really good.

> Groups in this instance could be referring to local, regional or national government, non-governmental organisations (NGOs), e.g. Farm Africa, and local community groups.

Developed world: farming

1 Study Figure 1.

Describe **one** reason why some farmers in the developed world have chosen to **diversify** into new activities.   **(2 marks)**

....

....

....

....

....

....

....

Figure 1 An example of a farm-diversification scheme in the south of England.

> Read the question! This is about WHY some farmers diversify not HOW.

2 Using examples, explain **how** some farmers have chosen to diversify their sources of income.   **(8 marks + 3 marks SPaG)**

One key diversification strategy is specialist and local foods, including drinks. These are often premium products, attracting more profits. These are often premium products which add value and therefore attract more profits. An example of this is ....

....

....

....

....

....

....

....

....

....

....

....

....

....

Developing world: farming

HIGHER

B

1 Study Figure 1.

Figure 1 A fair trade box of tea bags.

Describe **two** ways in which farmers in the developing world may benefit from fair trade schemes. **(4 marks)**

▷ **Guided** ▷

1 One of the main ideas behind fair trade is to reduce poverty. Farmers are paid a better price than they would be in the normal market / cash-crop scheme.

2 ....

....

....

....

FOUNDN

E/D

2 Describe **one** benefit of using intermediate technology. **(2 marks)**

....

....

....

HIGHER

B

3 Describe **one** way in which farmers in the developing world can improve **water supplies** to raise yields. **(2 marks)**

....

....

....

....

> It's always a good idea in these types of questions to use specific examples you have studied rather than a vague, general idea.

Unit 3

Unit 3 will test key geographical ideas that are embedded into the whole of your geography GCSE course. It is very different to Units 1 and 2 which are based on recall (as well as knowledge and understanding). This exam will also test application – i.e. transferable geographical skills.

Thinking like a geographer

This exam will make you think like a geographer – make geographical connections and links, especially between the core topic areas found in Units 1 and 2.

> At the end of the paper you will need to come to a decision about possible solutions to a geographical problem. You will be presented with different options, and whilst there are no 'right' or 'wrong' options, you will need to explain / justify your choice.

Understanding the topic and handling different resources

The topic in this workbook is based around the devastating earthquake that struck the Caribbean island of Haiti in 2010.

* Haiti is a poor country with low levels of development;
* it has an extremely vulnerable population, especially in urban areas where there are high concentrations of people;
* it also lies in a multiple hazard zone and experiences hurricanes and landslides as well as earthquakes.

> This is a complex topic, with a number of interrelated geographical ideas and connections.

> The questions in this section will help you practice for your Unit 3 exam but may not be representative of a real exam paper.

Poverty and vulnerability

Information on the problem:

What should Haiti's pathway of recovery be following the 2010 earthquake? There are many ways in which development money could be spent to help the people and the country recover from this devastating natural disaster.

- Some people think that the key priority should be putting money into primary education.
- Other people think that water and sanitation must be improved first.
- The government of Haiti must make a decision about where to put aid money for the future.

Section 1 – Poverty and vulnerability in Haiti

Indicator	Haiti	World ranking (position)
Age 15 and over can read and write	52.9%	N/A
HIV / Aids deaths	7 900	29/153 (1 is worst)
Life expectancy	62.85 years	186/222 (222 is worst)
Access to drinking water	~50%	N/A
Access to a clean toilet	~20%	N/A
Urban population growth rate (average annual)*	3.7%	N/A
Median age	21	N/A
Population of Haiti	9.8 million	0.7 million in Port-au-Prince
Children under the age of 5 underweight	18.9%	38/132 (1 is worst)

Source: *CIA World Factbook. +water.org. Note data is latest available.

Figure 1a Background information about levels of development in Haiti.

Poverty is at the centre of many disasters in a country as poor as Haiti. An earthquake like this (or a hurricane) does not cause the same damage in other countries, even those nearby like Dominican Republic, Jamaica and Cuba. They are hit by hurricanes every year and don't have any casualties, while Haiti has thousands. Haiti is very vulnerable – it is the badly constructed buildings that kill people, rather than the earthquake. Extensive deforestation and overall land and watershed degradation add to the country's vulnerability.

Figure 1b Haiti – a vulnerable population (adapted from a radio broadcast).

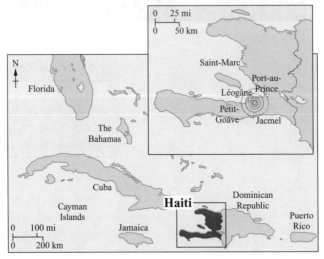

Figure 1c Map of regional location of Haiti.

Information on the problem

Study Section 1 and answer the following questions.

FOUNDN E/D

1 To the nearest one per cent, what was the proportion of underweight children in Haiti? **(1 mark)**

☐ **A** 18

☐ **B** 19

☐ **C** 20

☐ **D** 21

FOUNDN E/D

2 Approximately how many countries in the world have a **worse** life expectancy than Haiti? **(1 mark)**

☐ **A** 1–2

☐ **B** 5–10

☐ **C** 20–30

☐ **D** 30–40

FOUNDN D

3 Describe Haiti's geographical location. **(2 marks)**

....

....

....

....

> Use the map in Figure 1c plus your own geographical knowledge.

HIGHER C/B

4 What is meant the by idea of 'development'? **(2 marks)**

....

....

....

....

HIGHER B

5 Explain why low levels of development may lead to a population being vulnerable to different natural hazards. **(4 marks)**

Guided › Lack of money is the root cause of much of the vulnerability. It leads to poor housing

construction so people are vulnerable to earthquakes and cyclones. ....

....

....

....

....

....

Different levels of development in Haiti

Section 2 – Life expectancy and levels of development in Haiti

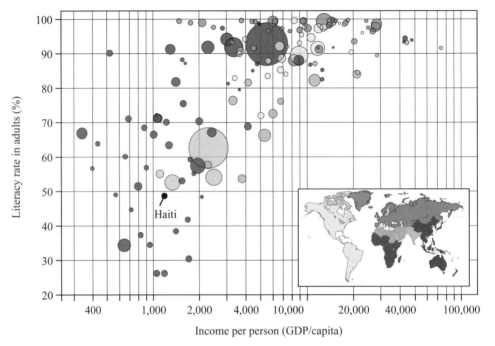

Figure 2a Literacy levels in Haiti compared to other parts of the world.

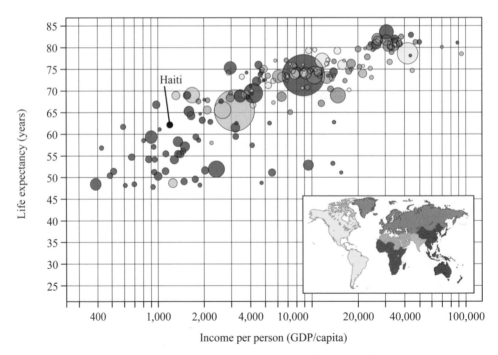

Figure 2b Life expectancy compared to other parts of the world.

Different levels of development in Haiti

You can also look back at the sources in Section 1.

Study Section 2 and answer the following questions.

FOUNDN
D

1 Describe Haiti's level of development in comparison with the rest of the world.   **(2 marks)**

....

....

....

....

HIGHER
B

2 Comment on the life expectancy for the people of Haiti in comparison with the rest of the Americas shown in Figure 2b.   **(3 marks)**

....

....

....

....

....

....

The 'comment on' instruction requires you to use your own geographical knowledge and understanding as well as the resource provided.

HIGHER
B

3 Examine the range of factors contributing to low levels of development in Haiti.   **(4 marks)**

....

....

....

....

....

....

....

....

Use your knowledge from Units 1 and 2 to help with these types of questions.

Impacts and costs of the disaster

Section 3 – Fact file on the causes and impacts of the earthquake in Haiti, 2010

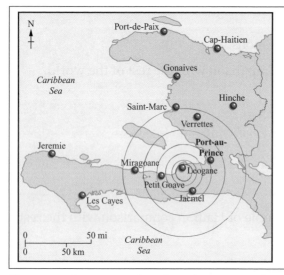

On 12 January 2010, just before 17:00, an earthquake of magnitude 7.0 on the Richter scale shook Haiti for 35 seconds. It was the most powerful earthquake to hit the country in 200 years. The earthquake was close to the surface (at a depth of 10 km) and its epicentre was close to the town of Léogâne, around 17 km south-west of the capital Port-au-Prince. The Port-au-Prince metropolitan area suffered extremely severe damage. Eighty per cent of the town of Léogâne was destroyed.

Figure 3a Summary of earthquake event and map showing the epicentre of the earthquake.

Seismologist 1
The earthquake caused extensive damage to buildings in the Port-au-Prince region and surrounding areas. Some 300 000 people died. The main port was also devastated during the earthquake, preventing the delivery of relief supplies.

Seismologist 2
The functioning of the government was seriously affected by the loss of people, records, and facilities. Numerous clinics, hospitals, police stations, churches, schools and universities were damaged.

Figure 3b Experts views on the impacts of the earthquake.

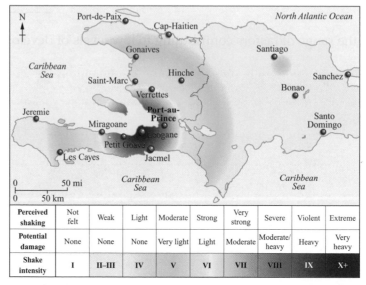

Perceived shaking	Not felt	Weak	Light	Moderate	Strong	Very strong	Severe	Violent	Extreme
Potential damage	None	None	None	Very light	Light	Moderate	Moderate/heavy	Heavy	Very heavy
Shake intensity	I	II–III	IV	V	VI	VII	VIII	IX	X+

Figure 3c Earthquake shake map.

Impacts and costs of the disaster

FOUNDN **D**

Study Section 3 and answer the following questions.

1 Seismologist 1 suggests that the earthquake had a serious impact on which vital **infrastructure**? **(1 mark)**

☐ **A** Schools

☐ **B** Port-au-Prince

☐ **C** Relief

☐ **D** The port

HIGHER **B**

2 Identify **two** pieces of evidence from Figures 3a and 3b to suggest that the earthquake had a significant impact on the country's capital, Port-au-Prince. **(2 marks)**

....

....

....

....

FOUNDN **D**

3 Describe the distribution of the most intense shaking in Figures 3a and 3c. **(2 marks)**

....

....

....

....

HIGHER **B/A**

4 Explain **one** piece of evidence from Sections 1 and 2 that links the geography of the earthquake to its impacts. **(3 marks)**

Guided ⟩ The epicentre of the earthquake was only 17 km from the capital (Figure 3a), meaning that there was a very large and vulnerable population close by with no time to evacuate. ....

....

....

....

....

....

....

> When asked for evidence, it's a good idea to refer to specific resources and / or pages from the resources provided. You can also make assumptions, e.g. 'poorly constructed houses' if this makes geographical sense.

Aid and development

Section 4 – Aid and recovery in Haiti following the earthquake

	Estimated needs (millions of dollars US)				
	2010	**2011**	**2012**	**2013**	**Total**
Recovery	1606	863	398	75	2942
Reconstruction	2912	1964	712	444	6032
Total	**4518**	**2827**	**1110**	**519**	**8973**

Figure 4a Estimated needs and recovery costs in Haiti, 2010–2013. Sourced from 'Haiti PDNA 2010', a World Bank publication.

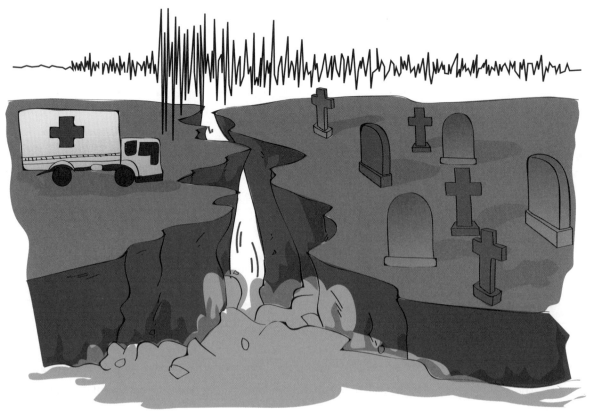

Figure 4b Cartoon about aid in Haiti.

Aid and development

Study Section 4 and answer the following questions.

FOUNDN

D/C

Guided

1 Suggest why there is a decrease in the reconstruction costs between 2010 and 2013. **(3 marks)**

Initially money will need to be spent on rebuilding infrastructure and housing ....

....

....

....

....

....

> It's a good idea to use data from the figures when responding to questions such as this. It may be a good idea to separate recovery and reconstruction. You should also draw on your own knowledge and understanding from Unit 1 when answering this question.

FOUNDN

F/E

2 Which of the following provides the best interpretation of the cartoon in Figure 4b? **(1 mark)**

☐ **A** Haiti is an earthquake disaster-zone waiting to happen.

☐ **B** Earthquakes produce different types of seismic shock wave patterns.

☐ **C** Relief from earthquakes cannot easily work so the result is many deaths.

☐ **D** Earthquakes kill lots of people as a result of poor infrastructure.

HIGHER

C/B

Guided

3 Suggest what might be the differences between **recovery** and **reconstruction** costs. **(2 marks)**

Recovery costs are often associated with dealing with the immediate after effects on the

disaster and they include medical treatments, emergency shelters, etc. ....

....

....

HIGHER

B

4 Outline the main message in the cartoon, Figure 4b. **(3 marks)**

....

....

....

....

....

....

Haiti's water and education status

Section 5 – Improving the water and primary education in Haiti

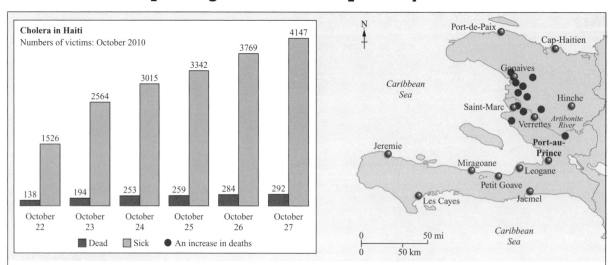

Cholera in Haiti
Numbers of victims: October 2010

138	194	253	259	284	292
1526	2564	3015	3342	3769	4147
October 22	October 23	October 24	October 25	October 26	October 27

■ Dead   ■ Sick   ● An increase in deaths

- Cholera had not been seen in Haiti for over 100 years, but there was an outbreak following the 2010 earthquake.
- The International Red Cross say that clean water distribution and hygiene promotion are key priorities.
- The health infrastructure was severely compromised by the earthquake, making the problems of treatment worse.

Figure 5a Importance of clean water in relation to cholera and cholera outbreaks following the earthquake.

In 2009, Water.org announced that it intended to help reach 50 000 Haitians with clean water and sanitation over the next three years, using a variety of solutions:
- Support local-level integrated water resources management, in particular the maintenance of drinking water quality and quantity.
- Improve access to sanitation and to safe, sustainable water for poor and vulnerable groups.
- Reduce the number of water- and sanitation-related diseases.
- Promote good hygiene practices.

Figure 5b Water.org clean water solutions and strategies, 2009.

- Haiti faces ongoing pressures of high population growth (more children to educate), high illiteracy rates and low primary education completion rates.
- A 2006 World Bank analysis reported that for Haiti to escape its poverty-conflict trap, it must invest heavily in its education sector for the 21st century – namely, the creation of a free, public and universal school system.
- In 2012 Haiti was ranked 161st out of 187 countries in the world for education spending.

Figure 5c Problems with primary education in Haiti.

Haiti's water and education status

Study Section 5 in the resource booklet and answer the following questions.

FOUNDN

C

Guided

1 Suggest reasons why Haiti needs to improve the quality of its water supplies.   **(6 marks)**

There is evidence that poor water supplies have caused recent outbreaks of cholera following the 2010 earthquake (e.g. information in Figure 5a). This will reduce people's ability to carry out economic work and activity, thereby further hampering the country's recovery. ....

....

....

....

....

....

....

HIGHER

A

2 Explain the main challenges associated with education in Haiti.   **(6 marks)**

....

....

....

....

....

....

....

....

....

HIGHER

A

Guided

3 Explain how climate change could further add stress to Haiti's water supplies.   **(6 marks)**

Climate change may lead to less reliable rainfall and rainfall which is lower or higher intensity at certain times of the year. Climate change may also alter the frequency of some natural hazards, e.g. cyclones or landslides, which again may affect the infrastructure which supplies clean water or takes away waste. ....

....

....

....

....

....

Use your knowledge from Unit 1, Topic 2 (Changing climate) to bring together ideas to help answer this question.

Making decisions

Foundation level

1 Study the **three** alternative options for aid to Haiti shown below.

> Option 1: Improve primary education.
>
> Option 2: Improve water and sanitation.
>
> Option 3: Improve both education and water / sanitation.

Select **one** option you think would be best for the people and the future of Haiti.

• Explain the advantages and disadvantages of this option for Haiti and its people. Use information from Units 1 and 2 and your own knowledge to support your answer.

(9 marks + 3 marks SPaG)

> **Guided**

Chosen option: |

Education is a very important aspect of development. Currently levels of literacy are very low in Haiti (Figure 2a). In order for people (a country's population resource) to have good jobs and be successful, they need a good primary education. ....

....

....

....

....

....

....

....

....

....

....

....

....

....

....

> This is going to be one of the most challenging aspects of your GCSE Geography exam, both in terms of the demand of the question and length of answer needed. Make sure your answer is clear and well-structured. Try and link and explain your points, and use geographical terminology. Also, be sure to check your spelling, punctuation and grammar.

Making decisions

Higher level

2 Study the **four** alternative options for aid to Haiti shown below.

> Option 1: Improve primary education.
>
> Option 2: Improve water and sanitation.
>
> Option 3: Improve the islands port infrastructure.
>
> Option 4: Improve education, water / sanitation and port infrastructure.

Select **one** option you think would be the best for the people and the future of Haiti.
- Explain the advantages and disadvantages of this option for Haiti and its people.
- Use information from Units 1 and 2 and your own knowledge to support your answer.

(12 marks + 3 marks SPaG)

Guided

Chosen option: 2

The country has a growing urban population (Figure 1a shows 3.7 per cent) therefore there is going to be an increasingly stressed water supply and pressures on sanitation. This would mean that option 2 would be the best option. The issues may be worsened by climate change which could lead to less reliable and less predictable rainfall. ....

....

....

....

....

....

....

....

....

....

....

....

....

....

....

....

....

....

> Remember, 3 marks are available for SPaG, so make sure you allow yourself time to read back through your answer. Check that your spelling, punctuation and grammar are really good and that your answer has a clear structure.

Unit 1 Dynamic Planet

The questions listed below will make up a practice paper. This will help you practice what you have learned, but may not be representative of a real exam paper. The pages on which these questions appear are listed below.

> Time: 1 hour 15 minutes
>
> You need to answer **all** questions in **Section A**.
>
> In **Section B**, answer **either** question 5 **or** 6.
>
> In **Section C**, answer **either** question 7 **or** 8.

Section A

Answer **all** questions in this section.

1. Restless Earth
1	Page 2, question 2	*(2 marks)*
2	Page 3, question 1	*(2 marks)*
3	Page 5, question 2	*(4 marks)*
4	Page 2, question 3	*(4 marks)*

Total for question 1: 12 marks

2. Changing Climate
1	Page 7, question 1	*(1 mark)*
2	Page 9, question 1	*(1 mark)*
3	Page 9, question 2	*(2 marks)*
4	Page 10, question 2	*(4 marks)*
5	Page 8, question 3	*(4 marks)*

Total for question 2: 12 marks

3. Battle for the Biosphere
1	Page 13, question 1	*(1 mark)*
2	Page 16, question 1	*(1 mark)*
3	Page 17, question 2	*(2 marks)*
4	Page 18, question 1	*(4 marks)*
5	Page 15, question 2	*(4 marks)*

Total for question 3: 12 marks

4. Water World
1	Page 19, question 1	*(1 mark)*
2	Page 23, question 1	*(1 mark)*
3	Page 21, question 1	*(2 marks)*
4	Page 21, question 2	*(4 marks)*
5	Page 22, question 2	*(4 marks)*

Total for question 4: 12 marks

Total for Section A: 48 marks

Section B

Answer **one** question in this section.

5. Coastal Change and Conflict
1	Page 28, question 1	*(1 mark)*
2(a)	Page 29, question 1(a)	*(1 marks)*
2(b)	Page 29, question 1(b)	*(2 marks)*
3	Page 25, question 2	*(2 marks)*
4	Page 28, question 3	*(6 marks + SPaG: 3 marks)*

Total for question 5: 12 marks + SPaG: 3 marks

6. River Processes and Pressures
1	Page 34, question 1	*(1 mark)*
2	Page 34, question 3	*(2 marks)*
3	Page 35, question 1	*(3 marks)*
4	Page 31, question 3	*(6 marks + SPaG: 3 marks)*

Total for question 6: 12 marks + SPaG: 3 marks

Total for Section B: 15 marks

Section C

Answer **one** question in this section.

7. Oceans on the Edge
1	Page 41, question 1	*(2 marks)*
2	Page 38, question 2	*(2 marks)*
3	Page 40, question 1	*(2 marks)*
4	Page 42, question 1	*(6 marks + SPaG: 3 marks)*

Total for question 7: 12 marks + SPaG: 3 marks

8. Extreme Environments
1	Page 43, question 3	*(2 marks)*
2	Page 45, question 1	*(2 marks)*
3	Page 45, question 3	*(2 marks)*
4	Page 46, question 2	*(6 marks + SPaG: 3 marks)*

Total for question 8: 12 marks + SPaG: 3 marks

Total for Section C: 15 marks

Total for Unit 1 Foundation paper: 78 marks

Unit 2 People and the Planet

The questions listed below will make up a practice paper. This will help you practice what you have learned, but may not be representative of a real exam paper. The pages on which these questions appear are listed below.

Time: 1 hour 15 minutes

You need to answer **all** questions in **Section A**.
In **Section B**, answer either question 5 **or** 6.
In **Section C**, answer either question 7 **or** 8.

Section A

Answer **all** questions in this section.

1. Population Dynamics
1	Page 50, question 1	*(1 mark)*
2	Page 50, question 2	*(1 mark)*
3	Page 49, question 4	*(2 marks)*
4	Page 51, question 3	*(4 marks)*
5	Page 54, question 3	*(4 marks)*

Total for question 1: 12 marks

2. Consuming Resources
1	Page 58, question 1	*(1 mark)*
2	Page 60, question 1	*(1 mark)*
3	Page 58, question 3	*(2 marks)*
4	Page 59, question 2	*(4 marks)*
5	Page 57, question 2	*(4 marks)*

Total for question 2: 12 marks

3. Globalisation
1	Page 64, question 1	*(1 mark)*
2	Page 66, question 1	*(1 mark)*
3	Page 65, question 1	*(2 marks)*
4	Page 63, question 2	*(4 marks)*
5	Page 66, question 2	*(4 marks)*

Total for question 3: 12 marks

4. Development Dilemmas
1	Page 71, question 1	*(2 marks)*
2	Page 67, question 3	*(2 marks)*
3	Page 71, question 3	*(4 marks)*
4	Page 70, question 2	*(4 marks)*

Total for question 4: 12 marks

Total for Section A: 48 marks

Section B

Answer **one** question in this section.

5. The Changing Economy of the UK
1	Page 73, question 1	*(1 mark)*
2	Page 74, question 1	*(1 mark)*
3	Page 77, question 1	*(2 marks)*
4	Page 78, question 2	*(2 marks)*
5	Page 77, question 3	*(6 marks + SPaG: 3 marks)*

Total for question 5: 12 marks + SPaG: 3 marks

6. Changing Settlements in the UK
1	Page 83, question 1	*(2 marks)*
2	Page 79, question 1	*(2 marks)*
3	Page 80, question 3	*(2 marks)*
4	Page 81, question 3	*(6 marks + SPaG: 3 marks)*

Total for question 6: 12 marks + SPaG: 3 marks

Total for Section B: 15 marks

Section C

Answer **one** question in this section.

7. The Challenges of an Urban World
1	Page 87, question 1	*(2 marks)*
2	Page 85, question 4	*(2 marks)*
3	Page 88, question 1	*(2 marks)*
4	Page 86, question 3	*(6 marks + SPaG: 3 marks)*

Total for question 7: 12 marks + SPaG: 3 marks

8. The Challenges of a Rural World
1	Page 95 , question 1	*(2 marks)*
2	Page 91, question 3	*(2 marks)*
3	Page 96, question 2	*(2 marks)*
4	Page 93, question 2	*(6 marks + SPaG: 3 marks)*

Total for question 8: 12 marks + SPaG: 3 marks

Total for Section C: 15 marks

Total for Unit 2 Foundation paper: 78 marks

Unit 1 Dynamic Planet

The questions listed below will make up a practice paper. This will help you practice what you have learned, but may not be representative of a real exam paper. The pages on which these questions appear are listed below.

> Time: 1 hour 15 minutes
>
> You need to answer **all** questions in **Section A**.
> In **Section B**, answer **either** question 5 **or** 6.
> In **Section C**, answer **either** question 7 **or** 8.

Section A

Answer **all** questions in this section.

1. Restless Earth
1	Page 2, question 1	*(2 marks)*
2	Page 4, question 2	*(4 marks)*
3	Page 5, question 3	*(6 marks)*

Total for question 1: 12 marks

2. Changing Climate
1	Page 7, question 2	*(2 marks)*
2	Page 9, question 3	*(4 marks)*
3	Page 11, question 2	*(6 marks)*

Total for question 2: 12 marks

3. Battle for the Biosphere
1	Page 13, question 2	*(2 marks)*
2	Page 13, question 3	*(4 marks)*
3	Page 14, question 4	*(6 marks)*

Total for question 3: 12 marks

4. Water World
1	Page 19, question 2	*(2 marks)*
2	Page 24, question 2	*(4 marks)*
3	Page 24, question 3	*(6 marks)*

Total for question 4: 12 marks

Total for Section A: 48 marks

Section B

Answer **one** question in this section.

5. Coastal Change and Conflict
1	Page 25, question 1	*(2 marks)*
2	Page 28, question 2	*(2 marks)*
3	Page 29, question 3	*(8 marks + SPaG: 3 marks)*

Total for question 5: 12 marks + SPaG: 3 marks

6. River Processes and Pressures
1	Page 35, question 2	*(2 marks)*
2	Page 36, question 3	*(2 marks)*
3	Page 33, question 3	*(8 marks + SPaG: 3 marks)*

Total for question 6: 12 marks + SPaG: 3 marks

Total for Section B: 15 marks

Section C

Answer **one** question in this section.

7. Oceans on the Edge
1	Page 38 , question 1	*(2 marks)*
2	Page 37, question 3	*(2 marks)*
3	Page 40, question 2	*(8 marks + SPaG: 3 marks)*

Total for question 7: 12 marks + SPaG: 3 marks

8. Extreme Environments
1	Page 43 , question 1	*(2 marks)*
2	Page 44, question 2	*(2 marks)*
3	Page 46, question 1	*(8 marks + SPaG: 3 marks)*

Total for question 8: 12 marks + SPaG: 3 marks

Total for Section C: 15 marks

Total for Unit 1 Higher paper: 78 marks

Unit 2 People and the Planet

The questions listed below will make up a practice paper. This will help you practice what you have learned, but may not be representative of a real exam paper. The pages on which these questions appear are listed below.

Time: 1 hour 15 minutes

You need to answer **all** questions in **Section A**.
In **Section B**, answer either question 5 **or** 6.
In **Section C**, answer either question 7 **or** 8.

Section A

Answer **all** questions in this section.

1. Population Dynamics
1	Page 51, question 1	*(2 marks)*
2	Page 51, question 2	*(4 marks)*
3	Page 53, question 2	*(6 marks)*

Total for question 1: 12 marks

2. Consuming Resources
1	Page 57, question 1	*(2 marks)*
2	Page 56, question 2	*(4 marks)*
3	Page 59, question 3	*(6 marks)*

Total for question 2: 12 marks

3. Globalisation
1	Page 61, question 2	*(2 marks)*
2	Page 63, question 3	*(4 marks)*
3	Page 65, question 2	*(6 marks)*

Total for question 3: 12 marks

4. Development Dilemmas
1	Page 67, question 1	*(2 marks)*
2	Page 70, question 1	*(4 marks)*
3	Page 69, question 3	*(6 marks)*

Total for question 4: 12 marks

Total for Section A: 48 marks

Section B

Answer **one** question in this section.

5. The Changing Economy of the UK
1	Page 74, question 2	*(2 marks)*
2	Page 77, question 2	*(2 marks)*
3	Page 75, question 3	*(8 marks + SPaG: 3 marks)*

Total for question 5: 12 marks + SPaG: 3 marks

6. Changing Settlements in the UK
1	Page 80, question 2	*(2 marks)*
2	Page 79, question 3	*(2 marks)*
3	Page 82, question 2	*(8 marks + SPaG: 3 marks)*

Total for question 6: 12 marks + SPaG: 3 marks)

Total for Section B: 15 marks

Section C

Answer **one** question in this section.

7. The Challenges of an Urban World
1	Page 85 , question 2	*(2 marks)*
2	Page 88, question 2	*(2 marks)*
3	Page 90, question 2	*(8 marks + SPaG: 3 marks)*

Total for question 7: 12 marks + SPaG: 3 marks

8. The Challenges of a Rural World
1	Page 92 , question 2	*(2 marks)*
2	Page 92, question 3	*(2 marks)*
3	Page 93, question 1	*(8 marks + SPaG: 3 marks)*

Total for question 8: 12 marks + SPaG: 3 marks

Total for Section C: 15 marks

Total for Unit 2 Higher paper: 78 marks

Answers

UNIT 1: DYNAMIC PLANET

Restless earth

1. Moving tectonic plates

1 D

2 Continental crust is normally made from granite and oceanic is normally basaltic.

3 As heat rises from the core, it releases convection currents in the form of cells. These currents are strong enough to move tectonic plates.

2. Plate boundaries, volcanoes and earthquakes

1 1 Found in lines (linear distribution).
   2 Mostly associated with the main plate boundaries around the Pacific Ocean.

2 Any one of the following.
   - Destructive boundaries – plates push together and the oceanic plate is subducted beneath the continental plate. Energy is released as earthquakes. Rising magma (oceanic) can create volcanoes on the continental crust side.
   - Constructive boundaries – plates forced apart by basaltic material which creates new crust. As the crust is pulled apart, it creates fissures and faults where molten magma can reach the surface forming volcanoes.
   - Collision boundaries – continental plates are pushed together buckling the crust and creating fold mountains.
   - Conservative boundaries – plates slide past one another causing earthquakes because of friction. Because there is no plate being created or destroyed, there is no magma rising to the surface and therefore no volcanoes.

3 Destructive plate boundaries are where oceanic and continental plates meet and the oceanic plate subducts under the continental one. The collision of the plates buckles the continental plate. The friction is very great which causes earthquakes. These can often be shallow. Earthquakes of up to magnitude 9.0 can occur and sometimes tsunamis, e.g. Japan 2011.

3. Volcanic and earthquake hazards

1 Any two from: strength and magnitude; depth of focus (whether it is close to the ground); location (i.e. earthquakes in remote rural areas will kill fewer people than those in built-up urban areas); vulnerability of the area (i.e. less developed countries will have fewer earthquake-proof homes etc.); time of day.

2 Physical factors, i.e. strength of the earthquake (magnitude), position relative to the coast as well as the depth of the epicentre of the earthquake.

3 The photo shows that the ground shook during the earthquake causing structural damage to buildings and break-up of the ground surface.

4. Managing earthquake and volcanic hazards

1 'Long term' is months or even years. 'Relief' relates to actions to help people and economies in that timescale, e.g. rebuilding houses, roads and infrastructure. Also helping to set up water supplies and farming practices again.

2 1 Aseismic buildings that are built to withstand earthquakes, e.g. reinforced structure with damping.
   2 Carry out practice evacuations so that impacts on people are minimised. Earthquake drills.

3 Aseismic buildings which move with the ground. In buildings, cross bracing and steel construction are used to reinforce the structure, e.g. Taipei 101 in Taiwan. Strong steel frames which prevent cracking. Double-glazed windows which add strength and also do not break easily when stressed. Some tall buildings have dampers which act like a pendulum counteracting building sway. Internally, heavy appliances are designed to be fixed to the floor / wall to prevent them moving. Reinforcing pipes for gas and water and using flexible joints to try to prevent them breaking and causing a fire or flooding.

5. Earthquake case study

1 'Primary' impacts are those impacts that happen straight away – examples would include buildings collapsing, people being injured / killed by falling structures, destruction of transport infrastructure (which complicates relief measures), etc.

2 For Haiti there were a number of people impacts, e.g. deaths (300 000), injuries, loss of homes, loss of livelihood, etc. The earthquake greatly damaged property (poorly constructed) because it was shallow and the epicentre was close to towns and the capital, Port-au-Prince.

3 Primary impacts for the 2010 Haiti earthquake included the death of around 300 000 people and many more injured and trapped; devastation of property – people's homes, schools, hospitals, businesses, etc. were damaged or destroyed. Secondary impacts included disease (cholera), killing thousands more due to poor living conditions in relief camps; loss of businesses meant people could not earn a living; disruption / difficulty in travelling; long-term economic disaster and problems for the country.

6. Volcanic eruptions

1 Any two from: lava flow destroying / damaging farmland, wildlife habitats; damaging property; cutting off rivers so endangering water supply; poisonous gases, ash and rock causing human and animal injuries and deaths.

2 Stopping of flights was a significant economic impact costing airlines and businesses millions of pounds in lost business and compensation. There was a significant social impact as flights could not be taken and families were disconnected; it was also at the start of a school term so overseas children could not get back to school!

3 Mt St Helens (1980). The newly established USGS developed a warning system (volcano watch). They monitored the seismic activity of the volcano and used computer models to try to predict when it would erupt and the extent of the volcano's damage. People were then warned about the risks and evacuated from their homes (Mt St Helens Contingency Plan).

Changing climate

7. Past climate change

1 B

2 Overall, a gradual increase from a mean of around 13.8°C in 1860 to 14.6°C in 2010. A fairly even rise until 1950 when there is a stepped increase and the rate of increase becomes greater.

8. The impact of climate change

1 See Guided for impact 1.
   Impact 2: during the Pleistocene ice ages when temperatures were much colder, megafauna (big animals such as woolly mammoths) lived in the UK. When temperatures warmed up, these animals became extinct.

2 Any two from: solar output, orbital changes of the Earth, major volcanic eruptions, sunspots.

3 People: changes in climate meant that people may have had to have grown new crops or historically hunted different animals due to their changed distribution.
   Ecosystems: some species have not been able to adapt and have therefore become extinct, e.g. the woolly mammoth at the end of the last ice age 8–10 000 years ago, which threatened the whole ecosystem.

9. Present and future climate change

1 C

2 Burning / combustion of fossil fuels, releasing CO_2 into the atmosphere (that was previously 'locked-up', e.g. as coal).

3 1 Farming intensification – cows releasing methane.
   2 Nitrous oxides – from jet aircraft engines.

10. Climate change challenges

1 People's houses, especially those close to the floodplain, are much more at risk from flooding, putting property and possessions at risk. This can also create a problem with insurance cover.

2 See Guided. Other problems are salinisation of farmland, meaning that it is very difficult to grow crops. There may also be a lack of space for building new houses and developing communities in areas where land is at risk as much of the country is low-lying.

11. Climate change in the UK

1 See Guided for positive impact.
   Negative impact: likelihood of increased droughts in some parts of the UK, e.g. SE England, coupled with less predictable rainfall patterns leading to more flooding.

2 There are a number of environmental challenges. These may include: in the summer, drought and water shortages could become more common, especially in the south; farmers may have to change their crops and crop yields could decrease through changing rainfall patterns; some plants and animals may struggle to survive with their existing patterns of

distribution; fish and marine ecosystems may also be
affected by rising water temperatures, causing loss of habitat
and migration.

12. Climate change in Bangladesh
1 Any two from: more disease; more conflicts between people;
more flooding and more cyclones causing damage to
property and injury / death etc.
2 See Guided. Another climate change impact will be a
possible increase in the frequency and intensity of cyclones,
which bring flooding and storm damage to people's
property and crops and possible injury / death.

Battle for the biosphere
13. Distribution of biomes
1 C
2 Tropical rainforests are found in South and Central America,
Africa and parts of SE Asia. Their distribution is closely tied
to within a few degrees north and south of the equator.
3 See Guided. Local factors also include aspect, geology / soil
type, drainage and altitude, which can all affect the type and
distribution of plants and animals within a biome.

14. A life-support system
1 1 Regulating the composition of gases in the atmosphere.
   2 Maintaining soil health.
2 Trees / plants contribute oxygen to the atmosphere through
the process of photosynthesis.
3 Good 1: biomass for energy.
Good 2: genetic resources.
4 See Guided. Tropical rainforests also act as a biodiversity store
of medicines and an important genetic resource (gene pool).
The biosphere also directly provides food for people (e.g. fruits
and berries) and sometimes water (e.g. baobab tree). It also
provides grazing and fodder to animals that humans depend
on for food, drink and other products such as clothing.

15. Threats to the biosphere
1 Uncontrolled deforestation is a serious problem which
removes the important biomass store. It leaves landscapes
scarred and vulnerable to soil erosion.
2 See Guided for effect 1.
Effect 2: tropical forest removal destroys soil nutrients as the
recycling process is stopped when the organic matter from
the timber and leaves is no longer forming part of the system.
3 Commercial intensive farming negatively impacts on
ecosystems. Wheat farming in the American Prairies has
removed natural grassland where bison once grazed. Urban
sprawl also destroys ecosystems and encourages wildfires,
e.g. Los Angeles. Commercial logging of the Amazon and
other rainforests destroys trees without replanting them.
Mining causes huge damage both through the mining
process itself and the creation of transport links such as
roads to transport the mined materials, e.g. Grand Carajas
development programme.

16. Management of the biosphere
1 One from: WWF, Greenpeace, Local Wildlife Trusts, etc.
2 Managing the behaviour of people, i.e. prevent people from
starting fires, thereby protecting the plants and animals from
damage and ensuring their future survival.
3 Countries can get together to develop wildlife conservation
treaties, e.g. RAMSAR (1971) for wetlands and CITES (1973)
to protect endangered species. Treaties such as these,
however, are difficult to manage as there are often
conflicting interests, but they do provide a useful framework
for conservation so that local management can be carried out
more successfully.

17. Factors affecting biomes
1 Temperature is a very important factor – it controls the
length of growing seasons and the types of plants and
animals that are adapted to the temperature regime of a
particular location. Precipitation is also very important
because all plants need it to varying degrees, e.g. a forest
ecosystem with a large biomass needs considerable rainfall.
2 In the north and west of Brazil, mainly found between 0°
and 10° south – restricted to quite a narrow belt of latitude.
3 A system that includes all living organisms (biotic factors) in
an area as well as its physical environment (abiotic factors)
functioning together as a unit.

18. Biosphere management tensions
1 See Guided for method 1.
Method 2: Biodiversity Action Plans (BAPs) protect natural
vegetation (and therefore some animals) in Britain. These
BAPs came out of Rio de Janeiro in 1992.

2 Sustainable management is all about protecting the
ecosystem for future generations. It may also involve
schemes which educate people and help people out of
poverty. Local people become stakeholders. In parts of
southern Africa, for example, local people can make money
from organising game tourism for visitors but this creates a
number of tensions. Other tensions / challenges from
sustainable approaches are centred on costs and funding
and sometimes the illegal division of funds / aid to gangs.
Other problems are concerned with who should pay for
sustainable management approaches.

Water world
19. The hydrological cycle
1 A
2 See Guided. Groundwater – water moves through the rocks
in the ground, often between areas of high pressure to low
pressure.
3 The hydrological 'system' is a continuous circle and a
never-ending cycle between the sea, the land and the
atmosphere. Water cannot escape this system because it is
closed.

20. Climate and water supplies
1 Most probable rainfall is in the east and north, the least in
the centre south and west of the country.
2 1 Insufficient water in the UK for instance, may lead to
hosepipe bans which prevents people washing their cars
and watering their gardens with a hose.
   2 Drought is disastrous for people who rely on livestock,
such as nomadic farmers. It kills their stock and therefore
their food and their way to make money.

21. Threats to the hydrological cycle
1 People will not have sufficient water to drink in some of the
driest parts of the world. This could eventually lead to
sickness, and in extreme cases death.
2 Threats to a healthy hydrological cycle include polluting
rivers and lakes through industrial pollution or sewage
which harms the health of the water itself. The hydrological
cycle can also be harmed if it is interrupted or held up, for
example reservoirs and dams stop the natural flow of the
cycle. Also, deforestation increases run-off so water reaches
rivers and streams quicker than is natural.
3 See Guided. In addition to dam construction, over-
abstraction of groundwater is another serious problem. It
can lead to saltwater intrusion which is widespread along
the Mediterranean coastlines of Italy, Spain and Turkey,
where the demands of tourist resorts are the major cause of
over-abstraction.

22. Large-scale water management 1
1 The largest proportion of UK rivers (41 per cent) are
Moderate water quality, with 35 per cent being described as
Good. There are only 4 per cent Bad and 14 per cent Poor.
2 See Guided. Other impacts include the building of large
dams which can cause sedimentation and siltation. Release
of organic pollutants into water courses, e.g. illegal dumping
of sewage, is also a cause.
3 China's booming economy and massive population are
posing some difficult environmental challenges for a nation
of some 1.2 billion people. Industrial water pollution is one
of the most pressing issues. Riverside chemical and power
plants, along with paper, textile and food production
facilities, are a leading source of pollution of China's rivers
and lakes, an estimated 70 per cent of which are now
contaminated.
Agricultural pollution is similar – about 11.7 million litres of
organic pollutants are emitted into Chinese waters every
day, compared to 5.5 in the United States, 3.4 in Japan, 2.3 in
Germany, 3.2 in India and 0.6 in South Africa.

23. Large-scale water management 2
1 Sardar Sarovar, India.
2 See Guided for advantages.
Disadvantages: dams can interfere with logging, navigation
by boats and fish migration. There is also an inevitable loss
of farmland and villages.
3 See Guided. Hoover Dam. Benefits included regulation of
water to reduce flood risks, water availability and
hydroelectric power. Costs included the construction of the
dam which has been credited with causing the decline of
this estuarine ecosystem, as well as the large impact on
the Colorado River Delta. It stopped natural flooding. The
construction of the dam devastated the populations of native
fish in the river downstream from the dam.

24. Small-scale water management

1. A use of technology which is appropriate to an area, i.e. it can be maintained and repaired without special tools and machinery so it is going to be workable. It is appropriate to the geography of the area.
2. Any two from: tube wells with water quality checking (developed by Bristol University) so that communities know their water is safe to drink; lining wells to prevent contamination by sewage; use of hand pumps to bring water up from wells – more efficient and less chance of contamination than using buckets and ropes; rain barrels – collecting and storing water for use on crops etc.
3. See Guided. Rain barrels – store water when it rains for people to use directly for drinking, washing, etc. and to irrigate crops; NGOs or others digging wells and providing hand pumps; lining wells to prevent contamination; hand pumps so more water can be pumped from deep underground and this is less likely to be contaminated than using buckets and ropes.

Coastal change and conflict

25. Coastal landforms and erosion

1. Stumps are a result of coastal erosion and cliff retreat. Hydraulic action opens up cracks and creates caves / arches in the cliff. These collapse and eventually leave stacks and then these are eroded to become stumps.
2. For abrasion: bits of rock and sand in waves grind down cliff surfaces like sandpaper.
   For attrition: material carried by the sea becomes rounder and smaller as it collides with other particles.
   For hydraulic action: the force of the waves hitting cliffs forces air into gaps and crevices in the cliff face, causing the rock to break apart.
3. Cliffs are shaped through a combination of erosion and weathering – the breakdown of rocks caused by weather conditions. Soft rock, e.g. sand and soft clay, erodes easily to create gently sloping cliffs. Hard rock, e.g. chalk, is more resistant and erodes slowly to create generally steep cliffs. See Guided. Over time, the notch gets bigger and the overlying cliff collapses, causing the cliff line to retreat.

26. Coastal landforms and deposition

1. Beaches are made up of eroded material (sand, shingle and stones) that has been transported from elsewhere and deposited by the sea. Where there is a drop in coastal energy, often caused by a change in the direction of the coasts (or shelter from a headland), this material is deposited and it forms a beach. Beach material is built up by swash. Longshore drift – a zig-zag movement along the beach caused by the waves approaching the coast at an angle – can 'move' the beach along the coastline.
2. Spits form as a result of sediment being carried by longshore drift along a coastline. At a river mouth, longshore drift pushes sediment out into the river. It is deposited forming a long neck of sand and shingle which eventually grows into a spit.
3. There are two contrasting types of waves. Destructive (storm waves) are associated with winter. These plunging waves with strong backwash create a steep beach profile. Constructive (spilling waves) are more common in the summer. These are lower energy waves and have a strong swash which pushes material up the beach. They are associated with a gentler beach profile.

27. Geology of coasts

1. Coastlines where the geology alternates between strata (or bands) of hard rock and soft rock are called discordant coastlines, whereas a concordant coastline has the same type of rock along its length. Concordant coastlines tend to have fewer bays and headlands. Some areas have a mixture of concordant and discordant sections, e.g. Isle of Purbeck in Dorset, part of the Jurassic Coast.
2. Sub-aerial processes are land-based processes which alter the shape of a coastline. They are a combination of both weathering and mass movement.
3. Swanage, Dorset. The coastline near Swanage is formed from bands of hard and soft rock (discordant coast). Waves have eroded the soft clay to form bays, e.g. Swanage Bay, and the resistant limestone to form headlands either side, e.g. Foreland, with caves, arches and stacks (Old Harry Stacks). Powerful, destructive waves hitting the cliffs have eroded weaknesses in the limestone to form caves. Over time, caves either side of the headland have eroded back-to-back until they've broken through to form an arch.

Continual erosion has widened the base of the arch until the top has collapsed, forming an isolated pillar known as a stack. Constructive waves have helped to form Swanage Bay, carrying sediment in the swash and depositing it to form a sandy beach.

28. Factors affecting coastlines

1. D
2. Main impact will be sea level rise flooding some low-lying coastal areas around the world, some islands may disappear completely. Another possible impact is that storms will become more frequent, therefore meaning more flooding.
3. Factors include: number of joints and faults in the rocks – more joints mean the cliff will erode quicker; strength of the waves – if the fetch is wide the waves will be more powerful and therefore have a greater impact; frequent storms in some areas mean more erosion; if sea defences are present, this will slow down erosion.

29. Coastal management

1. (b) Rip-rap works by absorbing and deflecting the energy of waves before they reach the defended structure. The size and mass of the rip-rap material absorbs the impact energy of waves, while the gaps between the rocks trap and slow the flow of water, lessening its ability to erode soil or structures on the coast.
2. Hard defences are normally only used in a 'hold the line' situation, i.e. to protect important infrastructures / houses etc. Examples include the Bacton gas terminal in North Norfolk, where the rest of the coast has retreat or do nothing management.
3. The problem in Essex is that sea level rise is taking place and will probably get worse. The coast is becoming constricted and the sea walls don't protect any more. A suggestion is that at Abbotshall, the sea should be allowed to flood the land behind and re-establish old salt marshes. The gently sloping land would then absorb the energy of the flood waters. Essex Wildlife Trust has purchased the land to make it an inter-tidal area suitable for wildlife. This is an example of a sustainable approach to coastal management.

30. Rapid coastal retreat

1. Coastal retreat is an area of coast suffering net erosion, so that the land is lost to the sea.
2. Monetary costs of advancing or holding the coastline may be huge and may not work for a long period of time. Land may be seen as 'not worth protecting', for example, if few homes and businesses will be affected. 'Do nothing' may be the most sustainable way of dealing with cliff retreat.
3. See Guided. Other problems include loss of valuable farmland as well as coastal golf courses, e.g. parts of the Cromer golf course which is situated between Cromer and Overstrand. A consequence of coastal retreat is that it can cause a drop in property prices locally, and personal distress for those whose homes and businesses are affected.

River processes and pressures

31. River systems

1. A long profile is similar to a slice through the river from source to mouth – along its long section. Many rivers have a concave long profile.
2. Channel shape becomes wider and deeper further downstream as the river contains more water. Gradient tends to decrease downstream, with the steepest sections being found in the headwaters of the catchment, and it becoming much gentler in the middle and lower sections.
3. See Guided. There is also an increase in depth. In the River Horner, Exmoor, depth increases to around 0.4 m near the mouth, whereas it is shallow (0.1 m) nearer the source.

32. Processes shaping rivers

1. Attrition: the current smashes rocks and pebbles in the river into each other, causing them to break and become smaller, smoother and more rounded.
   Abrasion: rocks and pebbles wear away the banks and bed of the river.
   Corrosion: rocks and minerals are dissolved by the river water.
   Hydraulic action: the force of the water wears away banks and bed.
2. See Guided. Other processes of transportation include suspension, saltation and solution. Saltation is a skipping movement of pebbles along the river bed whereas the tiny particles are carried in suspension, within the water itself. Solution is dissolved chemicals moving in the water.

3   See Hints. Geology may control slope profiles, with steeper long profiles being common in areas of hard rock. Bands of hard rock may also create waterfalls. Hard rocks will tend to form larger pebbles. Slope processes such as weathering and mass movement are responsible for the break-up and subsequent movement of material from the valley sides.

33. Upper course landforms
1   **(b)** In upland areas, small streams begin to develop and erode the landscape. A stream cuts vertically downwards, into the landscape, cutting a small V-shaped valley. Vertical erosion continues to erode the valley in the upper sections. Some parts of the hills tend to stick out into the river valley, resulting in a staggered formation, 'interlocked' together a bit like the teeth of a zip.

2   There are several erosion processes which create the different parts of a waterfall. Firstly, as the stream / river flows over rock, softer rock is eroded faster than harder rock. The River Twiss in North Yorkshire flows over limestone which is hard rock but when it meets a band of mudstone, a soft rock, it erodes it much faster. This has created a large drop which is now 10 m high, called the Thornton Force Waterfall. After the water drops over the cliff it continues to erode, creating a plunge pool. The water then undercuts the cliff by eroding the weaker rock underneath. This leads to the overhanging limestone eventually collapsing and the waterfall retreating upstream.

34. Lower course landforms
1   D
2   Ox-bow lakes are formed from the continued erosion of meanders where there is a narrow neck between two adjacent ones. Eventually, further erosion (lateral) causes the neck to be breached. The river flows down the quickest route depositing material on its banks which eventually cut off the meander to create an ox-bow lake.
3   Levees are formed by the process of deposition. Every time the river floods, sediment and water leave the river channel. The largest sediment is deposited on the banks forming raised embankments called levees.

35. Causes and impact of flooding
1   All river flooding is basically caused by heavy rainfall or snow / ice melt, both of which increase the amount of water in a river. However, several factors can increase flood risk, such as urbanisation and building on floodplains. Deforestation may also be a factor in some areas plus intensive agriculture, especially when large areas of fields are left with bare soil meaning that overland flow can become a problem.
2   Warmer air (from a warming climate) can hold more moisture, leading to more rain in the UK; climate change may also lead to greater intensity of rainfall.
3   See Guided and Hint. Increasing development of housing and industry on the floodplain is another contributory factor. Impermeable surfaces combined with quick draining roofs means that water is quickly transferred into the river. Equally, such building may also lead to less predictable and localised surface water drainage. Another human activity which may increase flooding is deforestation – trees and other plants soak up water and help slow down run-off, so reducing the foliage means the water reaches the river quicker.

36. Managing river floods
1   **(b)** Placed next to river, a flood wall prevents water at bankfull spreading into a town for instance, by acting as an impermeable barrier.
2   River Skerne. Sustainable / soft engineering approaches tend to be cheaper and visually less intrusive. Benefits have included a large increase in amenity land for local residents, with a 30 per cent increase in birds and insects after one year following the scheme. Locals liked the changes, 82 per cent of people liked it. Costs have been the work done and the loss of agricultural land. There is still uncertainty as to whether the scheme will cope with a really large-magnitude flood. Hard defences may have been a more effective alternative (but more costly).
3   'Sustainable' means something that can continue without causing damage to the environment and without needing a lot of maintenance or cost. Sustainable flood defences tend to be lower cost and have less of an impact on the environment – often includes allowing flooding in some areas, e.g. upstream where the land is not built on.

Oceans on the edge
37. Threats to the ocean
1   Coral reefs: mostly located between the tropics – between 30 degrees north and south of the equator where sea temperatures are warm. Most are located close to the coasts of Asia, Australia and Central America. There are also quite a few in the Pacific Ocean.
    Mangrove swamps: mostly located between the tropics – between 30 degrees north and south of the equator – around coastlines of Asia, Central America, Africa and Australasia.
2   Coral reefs: one of the direct impacts of climate change on coral reefs is sea temperature increases. This leads to coral bleaching (loss of symbiotic algae) and stresses the coral. There are a number of other serious threats including overfishing and destructive fishing practices, careless tourism (boating, diving, collecting coral). Pollution is another global threat, especially urban and industrial waste, sewage and oil pollution.
    Mangrove swamps: similar threats to coral reefs in terms of overfishing, pollution and tourism. In addition, a huge threat is that mangroves are being removed, often for prawn aquaculture. Climate change may also impact on mangroves as sea temperatures rise – some species may not be able to survive, additionally as sea levels rise many mangroves may disappear.
3   Any one of: unique ecosystem worth protecting; conserving habitats for plants and fish; mangroves protect the coastline from storm surges and tsunamis; plants that grow there provide an important food supply for wildlife e.g. deer, snakes and crocodiles; also provide food for humans so their loss will negatively affect fishing industry; will negatively affect tourism; mangroves are a plant and therefore consume CO_2 and produce oxygen, so getting rid of them will add to climate change.

38. Ecosystem change
1   Any one of: pollution caused by litter, industrial waste, oil spills, etc. all damage and destroy part of marine ecosystems; human activities are contributing to climate change which threatens marine ecosystems as sea temperatures and sea levels will rise; clearance, for example, of mangroves for prawn farming; overfishing interferes with natural balance of the ecosystem; fertilisers from farming causes eutrophication in the seas where algae explodes and uses up most of the water's oxygen, causing the death of other species; siltation caused by deforestation buries plants so they die.
2   Any one of the following.
    • Coral reefs: industrial pollution from urban run-off; pollution from oil / fuel spills from boats; tourists collecting coral / damaging it due to carelessness; overfishing threatening extinction of some species.
    • Mangroves: clearance for prawn farming; damage from storms; overfishing; pollution from urban run-off and boats.
3   See Guided. Climate change is leading to an increase in sea temperatures globally and coral reef ecosystems are especially sensitive to changes in water temperature. When summer water temperatures are just a degree or two warmer than usual for a few weeks, a critical yet delicate symbiotic relationship breaks down and the algae are expelled, often leading to the coral's death. Overfishing disrupts the ecosystem; eutrophication from nitrates which wash into the sea; siltation caused for example by deforestation, buries marine plants causing them to die. Local fishermen, especially in some developing countries, have used dynamite to kill fish – this not only kills the fish but damages the local environment.

39. Pressure on the ecosystem
1   B
2   Eutrophication is caused by the run off of agricultural fertilisers – the main compounds in the fertilisers are nitrogen and potassium. See Hint and previous for more on eutrophication. Overfishing is a serious threat that can disrupt marine food webs in some instances. Anchovies and sardines, for example, are staple food for salmon, tuna, seals, terns, pelicans and many other species. In the marine food chain, the small fish are a link between bigger predators and the tiny floating plankton that absorb energy from the Sun, just as plants on land do. This also can lead to a breakdown in nutrient recycling.

40. Localised pressures
1   The Soufriere Marine Management Area was set up in the early 1990s to protect that area of the Caribbean. It was based around a coastal-zoning system.

2 Any suitable example could be used here.
Fish stocks in the North Sea. The EU Common Fisheries Policy has attempted to bring back fish stocks from catastrophically low levels. This has involved working with different stakeholders to try to manage the stock using a whole ecosystem approach. One of the most conflicting views is restricting the size of holes in nets so that young fish are not caught. Some fishermen also disagree with the quota management system and the issue of by-catch (although 'discard management' is being considered). Scientists want to set up no fishing marine reserves in the North Sea, but again there are conflicting views as these are expensive to establish, monitor and maintain.

3 Simply a way of describing the state of the marine ecosystem. A healthy ecosystem is one that can maintain or increase benefits (food and services) in the long term, without damaging the health of the marine web of life that provides these benefits.

41. Local sustainable management

1 An overall decrease between 1963 and 2003. Most significant decline was after 1968. There was actually an increase between 1963 and 1968, up to 270 000 tonnes.

2 See Guided. This is the Lamlash Bay Community Marine Protected Area, set up in 2008. This No Take Zone (NTZ) aims to protect seaweed beds and the regeneration of all marine life. It is hoped this NTZ will benefit the Lamlash economy by attracting scuba divers and by providing bigger and better catches for fishermen in the neighbouring fishing area.

3 Sustainability refers to the longer term, so in this instance marine sustainability is about preserving fish stocks so that they are available for future generations and not overfished (in an unsustainable way).

42. Global sustainable management

1 See Guided. There are, however, problems with the MPAs. Some are very small (1 km²) which makes them quite ineffective, and many are poorly managed. Also, they do not always protect the areas of highest biodiversity – the hotspots. Conservation International has a global marine programme. They are acting globally to raise awareness of marine issues and produce the Ocean Health Index. There are various examples of agreements, e.g. the Law of the Sea developed at UNCLOS. This covered a range of strategies including attempting to manage pollution and fisheries. It was finalised in 1994 and 40 per cent of the ocean (next to mainland) was placed under its protection. In 1974, the Helsinki Convention developed programmes to manage marine pollution. It led to a series of UNEP and regional action plans where states work together to clear up seas collectively, such as the Mediterranean.

2 See Guided. Areas vary from 79 km² up to a massive 80 000 km² in Norway. This is perhaps a reflection of how important they see marine life in terms of their economy (fishing) and well-being.

Extreme environments

43. Extreme climates: characteristics

1 Mainly found a few degrees north and south of the equator. Largest concentration in northern Africa and the Middle East, plus Australia. Some smaller deserts in USA, South America and parts of southern Africa.

2 D

3 Any two from: very cold; dry with little precipitation; most precipitation is snow; ground is either ice (glacial) or frozen soil (tundra); seasons where winters are very cold and few hours of sunlight and summers less cold and lots of hours of sunlight.

44. Why are extreme climates fragile?

1 C

2 Any suitable example such as, in Australia succulent species are capable of storing water in their fleshy leaves, stems or roots. They are adapted to living in areas of low water and are drought tolerant. Or, in tundra regions sedges and moss are adapted so they thrive in boggy conditions with little heat or sunlight.

3 Any suitable example such as, the red kangaroo feeds at dawn and dusk when the air is cooler and it hops which is an energy efficient way of travelling. Or, polar bears have very thick fur and layers of fat to survive the cold, and are coloured white to camouflage them against the snow, helping them to hunt.

45. People and extreme climates

1 Farmers base their agriculture on mountain grazing for the livestock in the summer and simultaneously cultivate grass

in the fields near the farms where they live, which could be used as winter feed for animals.

2 A – Masai: loose-fitting clothing which is light, made of cotton and allows circulation of air during the day to cool the body.
B – Inuit: warm clothing, furs which are heavy but durable and very good in extreme cold. Boots also keep the feet warm.

3 Any one from: steep roofs so snow slides off; triple glazing to keep the inside of buildings warm; houses on stilts where there is permafrost so they do not melt the frozen ground and destroy / damage the building.

46. Threats to extreme climates

1 See Guided. They are also using mobile technology to their advantage, e.g. Kenya M-Pesa is used to deliver cashless payments between people via mobile phones (e.g. Safricom). The environmental changes brought on by global climate change are having a devastating cultural and psychological effect on the Inuit. Traditions like hunting marine mammals, which depend on plentiful sea ice, are in decline.

2 For any hot arid or polar region several threats could be mentioned: pollution (See Guided); impacts of oil and gas development (e.g. in the Arctic) – tribes and indigenous peoples have voiced their opposition to offshore oil and gas activities that are being proposed; climate change puts environments at risk (e.g. for polar, the Mackenzie River Basin, north-western Canada, is especially at risk because extremely warm temperatures during the summer of 1995 sparked widespread wildfires forcing the evacuation of Tulita, a small Basin community).

47. Extreme climates: sustainable management

1 Ecotourism is intended as a low-impact and often small-scale alternative to standard commercial (mass) tourism. For the last 20 years, WWF, for example, has had an Arctic programme to facilitate codes of Arctic tourism. Local players have included tour operators, local government, residents and members of indigenous groups. They have together identified 10 important principals for environmentally and culturally appropriate tourism.

2 Any suitable examples are credited. See Guided. In Zambia, the community-based Evangelical Fellowship of Zambia (EFZ) trains people to use conservation farming and multi-cropping which makes their approaches much more sustainable for the future. The Arctic Co-operation Programme is designed to promote sustainable development in the Arctic. For example, supporting initiatives that improve education, enhance the skills of the people and strengthen the grass-roots level and the work done by organisations in the region.

48. Extreme climates: global management

1 Hot arid: climate change will increase aridity making it harder / impossible to grow crops or graze animals so people will find it harder to feed themselves or make a living; sand dunes could move and threaten more crops and grazing land. Polar: glaciers melting may lead to flooding, damaging people's homes / businesses / crops, etc. and will also stop tourism to those areas; species adapt to warmer areas – may impact on people living there; melting of permafrost will lead to flooding.

2 Polar and arid regions are special places and therefore international efforts have been in place for a long time to try to protect them. One of the earliest pieces of legislation was the 1961 Antarctic Treaty restricting commercial exploitation. The 1998 Protocol on Environmental Protection followed this, preventing new activities that may threaten the Arctic.
2006 was the International Year of Deserts and Desertification. The UN tried to promote the significance and specialness of deserts and their communities and to raise awareness about threats to its longer-term sustainability.

UNIT 2: PEOPLE AND THE PLANET

Population dynamics

49. World population growth

1 B

2 In stage 4, there is a gradual increase in the total population over time but at the beginning of Stage 5, the death rate per 1000 people overtakes the birth rate and the total population begins to decline.

3 Any one from: education has taught women about contraception and family planning; educating women means many have fewer children in order to pursue careers and / or put off having children until later in life so they end up

having fewer; advanced health care means fewer children are dying so there is less need to have more to start with.

4 Growing at an increasingly rapid rate (has a very steep, J-shaped curve).

50. Population and development
1 C
2 Ages 30–34.
3 See Guided. Pyramid B has low birth rates and death rates. There is an increasingly ageing population (dominance of women in the pyramid aged 75+). There tends to be effective birth control.

51. Population issues
1 A 'dependant' is someone who needs the support of the state and other people both financially and socially. Usually means children or elderly.
2 Youthful population is more pyramid shaped – wide base going up to a peak. Shows a high number of people under 30 years in age.
Ageing population is more 'tower shaped' with a narrow base. The top of the pyramid is equal or greater than the youthful part.
3 Problems of ageing population include: pension burden; more need for social care such as home help, meals on wheels; suitable housing; increased need for care homes; increased need for carers – either need to pay for someone or a family member could give up work to do it and therefore lose income; increased need for hospital beds and professional care for diseases affecting the elderly, such as dementia; feelings of loneliness and isolation of the elderly.

52. Managing populations
1 Sustainable level of population means the right amount of resources – food, jobs, houses, services – for the number of people.
2 Any one of: not enough resources for the population may lead to malnutrition and starvation; shortages of services such as schools and hospitals; shortage of housing; overcrowding leading to poor living conditions, disease etc.; not enough jobs means unemployment would be high, leading to poverty and a strain on benefits (if the country has them).
3 Governments try to manage populations so that there is a balance between resources, e.g. jobs and the number of people available to work for instance. In the 1950s, there was a catastrophic famine in China which killed millions of people. This is one reason why the Chinese government introduced its One Child Policy in 1978 to try to reduce the birth rate. Having more than one child is discouraged in China, and unauthorised births are punished by fines. Other places have experienced underpopulation and / or an ageing population, so encourage people to have more children. These countries, such as Singapore, also encourage migration to help with skills and labour shortages.

53. Pro- and anti-natal policies
1 A pro-natalist policy includes incentives to encourage people to have more children or larger families, i.e. to try to increase the birth rate.
2 See Guided. China's One Child Policy prevented 300 million births and therefore was a success in that respect. It also prevented another catastrophic famine. Disadvantages include the harshness of the policy, making people feel resentful and the 'one child' is often spoilt – 'Little Emperors'. There are now worries that China's rapidly growing economy will not have enough young workers to keep it going. There has also been widespread sex-selective abortion – because the preference is for boys there is now an imbalance of too many men and not enough women.
3 Pro-natalist policies such as good childcare, tax incentives; access to the best schools and nurseries etc. Alternatively, any policy which encourages migration is another attempt to increase population.

54. Migration policies
1 D and E
2 See Guided. Taking the test will give the person the practical knowledge they need to live in this country and to take part in society.
3 Reasons include: to fill labour shortages; to fill skills shortages; to help pay for services through taxes (especially in countries which have an ageing population); most immigrants are likely to be young and therefore likely to have children which will also help with underpopulation.

Consuming resources
55. Types of resources
1 Any one from: coal, oil or gas.
2 Any one from: they can be reused again and again and therefore there is an indefinite long-term supply of them; after initial set-up costs, carbon emissions are far less than from non-renewable sources.
3 Advantages: 'green' clean energy with no emissions, energy source will not run out; benefit locals, i.e. for individuals or communities (especially those who are off-grid). Disadvantages or conflicts: efficiencies may not be all that is claimed; can be harmful to birds; some people do not like the look of them and the noise they make.

56. Resource supply and use 1
1 Any two from: lifestyle and technology, e.g. more energy consumption as have more electrical goods, e.g. washing machines; higher car ownership due to higher levels of wealth; better developed travel infrastructure, e.g. road and rail uses more resources; generally higher levels of wealth mean more people can afford more fuel for heating and eating.
2 There is often a mismatch between supply and usage of energy in some counties and regions, i.e. the Middle East have a plentiful supply of oil for instance, other countries have a high demand and usage of oil e.g. the USA, so that can cause tension between them. Energy can be supplied in many ways, which may also create conflict and tension – for example, oil has to be transported on ships which may lead to pollution of the sea affecting some countries, gas pipelines have to be built across people's land, etc.
3 See Guided. Most growth in energy consumption will come from developing countries – as they develop they will use greater energy – more people will be able to afford cars, household appliances, computer technology, etc. Resource supply will continue to be unequal.

57. Resource supply and use 2
1 All of these renewables depend on the natural resources available – few places have access to geothermal energy (e.g. Iceland), similarly not all places get many hours of sunshine to use solar power, enough wind or are suitable for growing biofuels. More countries have access to rivers / lakes etc. to make use of hydroelectric power, hence the reason why it is the largest renewable.
2 Coal is consumed by a large number of countries round the globe, but China and the USA are by far the largest users. In total, there are about 40 countries which use it. Coal is also found in a large number of countries. China supplies 40 per cent of the world's coal and the USA about 16 per cent. UK is 0.3 per cent.

58. Consumption theories
1 A
2 Over time, both food required and food produced increase. At point X, food required increases beyond that of food produced, i.e. there will be a food deficit. Note the difference in the shape of the curves – linear vs exponential.
3 Boserup's theory is basically that population growth has a positive impact on resource supply – as resources start to run out, people innovate to cope with the problem.

59. Managing consumption
1 Thinking carefully about how resources are reused now to ensure that they have a sustainable future.
2 1 In the UK, cars are taxed according to their CO_2 emissions and engine size. The government makes the more polluting engines more expensive to tax and so manages petrol / diesel consumption and penalises users.
  2 In the UK, encouraging individuals to use energy more sustainably in the home by offering cheap home insulation grants and upgrades to more efficient boilers etc.
3 See Guided. The UK government has introduced a FIT (Feed-in Tariff) scheme to promote householders' use of green electricity generation. Some other governments have encouraged greener transport such as more cycling through cycle-to-work schemes and bike sharing, e.g. 'Boris Bikes' in London.

60. Potential of renewables

1 Can be reused and / or renewed in a reasonable timescale.
2 Alternative technologies, such as hydrogen, limit the reliance on fossil fuels, mainly oil. Other technologies, such as more fuel-efficient vehicles, duel fuel, hybrids, etc. may also offset resource shortages. Technology may mean that controversial tar sands oil and fracked gas also become commercially viable. Technology may allow people to use less energy at home, e.g. better dishwashers and washing machines etc.
3 See Hint. Technologies such as hydrogen transport, innovative buildings which use less carbon to build and do not waste energy. New technologies improve existing inventions – less energy needed for domestic appliances, for example. New technologies are also used to supply renewable energy and other energy sources, such as fracking and tar sands.

Globalisation

61. Changing employment patterns

1 See Guided for primary employment.
Quaternary: an increase in the number of people employed in this sector, from 0 to about 5 per cent.
2 Any one from: shift of manufacturing and linked jobs away from the home country to other more cost-effective locations; increasing urbanisation means there is more need for services in towns and cities to support the population – transport, finance, education, health care; better education means more people going to university and therefore work in tertiary and quaternary sectors increases.
3 Use Figure 1 to assist. This is a simple model which shows how the stages in employment sectors change as a country's economy grows. The model describes a linear sequence of predictable changes, i.e. economy based almost totally on agriculture, such as many countries in Sub-Saharan Africa (1); then a rise in industry when primary employment decreases and manufacturing increases (2), such as in China; then a rise in the service sector and a decrease in the secondary sector, such as in the UK (3). But it is a model based on a traditional idea of development in countries such as the UK. It may not work in the same way for some emerging economies today (e.g. China, India, Brazil, Thailand, etc.) who may bypass / merge some of the stages.

62. Employment sectors

1 In Latin America / Caribbean, there is an average of about 16 per cent employment in agriculture, whereas in Sub-Saharan Africa it is much higher, nearer 60 per cent employment. In Latin America, employment in services is the largest proportion, with women especially (around 80 per cent). This compares with 35 per cent in Africa. For both regions, industry accounts for the smallest proportion of employment.
2 See Guided. According to the Clark Fisher model, as a country develops the percentage of people working in different employment sectors change. At the lowest level of development (the pre-industrial stage), most people are employed in primary industries such as farming, fishing and mining. Many of these will be subsistence farmers / hunter-gatherers. As a country begins to develop, it reaches the industrial stage where primary sector employment decreases and secondary industries dominate. A few more people are also employed in the tertiary sector. As a country continues to develop, it reaches the post-industrial stage where tertiary sector employment becomes the largest sector and secondary declines. After the post-industrial stage, the tertiary sector still dominates but the quaternary sector begins to grow. As a country develops through all stages, a wider variety of employment becomes available.

63. Impact of globalisation

1 The answer will depend on your chosen organisation. Points might include:
Manufacturing, sales, marketing, location of headquarters, type of product, TNC status, customer base.
Coca-Cola is a transnational corporation (TNC). Its headquarters are in the USA. Manufacturing and sales outlets for the products of Coca-Cola are located in different places across the world. It invests in different countries of the world so it can expand into new markets and keep costs down.

2 Advantages might include: increased employment; increased availability of products and services; the development of new infrastructure, e.g. roads and rail to support the developments. Disadvantages might include: exploitation of labour; lower wages for most; few people can afford the new products; increased pollution from manufacturing operations.
3 Some landowners in developing countries have benefitted hugely as they have received vast amounts of money for the use of their land and the resources on it; most people across the world except for the poorest have benefitted from the increased number of products and services that globalisation has brought; improvements in education has benefitted most, except the most poor throughout the world; mostly men have seen the biggest benefits – mostly the men who own and run TNCs or own land etc.; increasing awareness; industrial workers – conditions have improved for all (but remain poor in many part of the developing world).

64. International trade and capital flows

1 A
2 One of the key processes in a 'shrinking world' is the role of technology, e.g. the internet, but also linked to better transport, allowing goods and people to travel more easily and quickly.
3 Technology has improved transport and communications – shrinking the world. Companies have grown bigger, expanded into new markets and therefore have overseas operations. There has been a growth of TNCs. Trade agreements have been set up to promote trade between different countries. There has also been more interest in the sharing of ideas, experiences and lifestyles of people and cultures. People can experience foods and other products not previously available in their countries – this has contributed to international trade.

65. TNCs: secondary sector

1 TNCs have operations in more than one country. They are usually capitalist enterprises, making money for shareholders. Usually the head office and design are based in developed countries and manufacturing in developing countries to take advantage of lower costs.
2 Any suitable TNC could be used, this is just an example. Toyota is based in Japan, but has operations around the world. Although based in Japan, Toyota produces most of its cars in its transplants in Georgetown, Kentucky, and Burnaston, Derbyshire. Toyota is a typical transnational corporation that understands that considerable gains can be made by locating manufacturing plants outside their country of origin. Toyota expanded to Europe in 1992 in order to achieve the benefits associated with establishing a manufacturing base in another country.
Transnational corporations search out locations for their plants outside their original country because they can overcome costs such as transport and duties on imports and exports. They can also access new markets more easily, e.g. European Union.
3 Mainly involves moving manufacturing operations to lower cost and / or more efficient locations where production can be outsourced, but has also happened to service sector work – outsourcing of back-office jobs. Typically a process linked to TNCs over the last 30 years.

66. TNCs: tertiary sector

1 Examples could include: Barclays (Bank), Tesco.
2 Any suitable example could be used. For example, for Tesco: the headquarters are located in the UK; it outsources many products from all over the world, such as flowers from Kenya; they have stores in many parts of the world (e.g. a large number in Asia). You should think about adding real examples and data where possible to develop your answer.
3 The shift has happened to make the operation more cost efficient because of lower staff and operational costs in developing countries; it is also helped by the fact that back-office operations, e.g. computer support / servers, can be moved to remote locations with skilled but lower cost workforce – due to the internet you no longer have to be in the same office or even country as your colleagues.

Development dilemmas

67. What is development?

1 Highest levels of GDP are found in the USA and Europe ($45 000). The lowest levels are in Sub-Saharan Africa. Generally, GDP is lower than average in SE Asia, especially India which has only $2000–4000. Australia has the highest GDP in the southern hemisphere.

2 A

3 GDP per capita means Gross Domestic Product per person – in other words the total wealth of a country divided by its population.

68. The development gap

1 C

2 The HDI is a composite index of a number of different variables, e.g. life expectancy, education and income. Countries are then ranked against each other: HDI near 1 = highest levels of development, low scores nearer 0.5 and below indicate lowest levels of development in a world context. This helps us to understand levels of development at a glance.

3 The development gap has become wider because rich countries tend to trade with each other and generate more wealth leaving other countries behind. Over time, you might expect the gap to be reduced, especially with more globalisation and better trade etc. Emerging economies, particularly countries such as Brazil, China and Indonesia, have become richer and so they are now more like the UK 20 years ago in terms of wealth. But some countries are still locked in poverty, with poorly run governments and corrupt leaders. They may also suffer natural hazards and disasters which can widen the gap.

69. Development

1 Corruption. TNCs and foreign investors will be reluctant to invest in corrupt countries / regimes as they will think it is difficult to do business there. Also, corruption often means that money from aid-targeting investment projects is used inappropriately for personal gain.

2 Any suitable example will be fine.
   Nigeria, for example, is a country rich in natural resources but it faces a number of development problems. Government leadership is one of the key issues along with corruption. The country has also been beset with civil war. There is also a poor educational system with poorly rated universities.

3 Any suitable example will be fine.
   Low levels of development in Kenya, for example, can be explained by a number of factors. There are several issues around corruption and bribery of officials. There are also problems with politics and power in terms of tribal dominance. Another barrier to development is water and drought management, especially in the north of the country – it is a cause of rising poverty. Like most Sub-Saharan countries, the road infrastructure is also very poor meaning that less inward investment has been attracted.

70. Theories of development

1 Rostow's theory tries to explain how a society develops over time from an economic perspective. It shows that to develop, countries have to meet pre-conditions, such as having educated people to do the work. Once countries have educated people and start saving and investing money, they develop fast. However, this theory assumes development is non-stop so doesn't take account of problems when economies slow down or go backwards.

2 Such theories have worked well in terms of describing development and changes in societies for countries such as the UK. But theories often assume that population changes are induced by industrial changes and increased wealth, without taking into account the role of social change in determining birth rates, e.g. the education of women. The theories may not explain the early fertility declines in much of Asia in the second half of the 20th century or the delays in fertility decline in parts of the Middle East.

71. Regional disparity

1 Big range of GDP within China – Guangdong is 80 times bigger than Tibet in terms of output ($838 vs $10 billion). Growth rates broadly similar 10–16.5 per cent.

2 The provinces with the largest populations have the greatest GDP, i.e. Guangdong has the largest population of 105 million people and the greatest GDP. The province with the smallest population, Tibet, has the smallest GDP. However, the pattern is not so straightforward where economic growth is concerned. The provinces with the biggest growth rate – Chongqing and Guizhou – have fairly small populations. Whereas the province with the largest population has the lowest growth rate, possibly because it has a relatively high GDP and therefore has less need for development.

3 Physical geography may play a part – very steep slopes and high altitude make agriculture and building infrastructure difficult so these become very isolated. Climate may also play a part – areas with little rainfall, for instance, may be disadvantaged. Other factors many include a historical dimension, e.g. which places were first settled and then had a catchment and geographical hinterland (sphere of influence). The government may also favour development in certain areas, e.g. coastal zones, and these have been given preferential treatment in China for example.

72. Types of development

1 These tend to be large-scale, centrally managed projects which are often very expensive.

2 Any appropriate example can be used.
   The Sardar Sarovar dam, for example, across the Namada River is nearing completion. The dam is one of India's most controversial dam projects and its environmental impact and net costs and benefits are widely debated. Local people have been negatively affected with 320 000 removed. Good quality farmland has been submerged – a problem for the farmers. Benefits include electricity for nearby cities and industries, plus canals which will irrigate farmland in areas such as Gujarat, providing a benefit for people in other states.

3 Top-down development projects tend to be very large scale. Bottom-up development projects tend to be smaller scale and locally organised and managed. They usually also have a shorter time frame of development and implementation. Top-down are usually more expensive.

The changing economy of the UK

73. Industrial change in the UK

1 C

2 Cheaper to buy coal from overseas where labour is cheaper. There was less demand for coal in the UK, e.g. in power stations, as they began to switch to cleaner alternatives, such as gas. Little political will to invest in improving performance of coal-fired power stations.

74. UK employment

1 D

2 Increase in part-time jobs may be due to people and businesses wanting a more flexible labour force, allowing women to work and look after children. Temporary jobs have grown due to recession and rise of contract work – again offering more flexibility.

3 Enables companies and individuals to save time and therefore money through the need for less travel which therefore maximises working time. These ways of working give flexibility so people work where and when they want and allow people to answer emails etc. on the move. Employees have to visit their offices much less often so potentially it is a greener way of operating. Also means that more people are able to live further away from their office / headquarters of work.

75. UK regions and employment

1 Any named industry and area could be used. For example, many financial services have chosen to locate in the south east of England because of its proximity to the capital city – centre of government and banking etc.; also, it has good communication links to Europe and the rest of the UK.

2 Any example region could be used. In the north east in the 1970s, many people were employed in primary industry (coal mining) and secondary industry (steel making, chemical and shipbuilding). This decreased rapidly in the 1980s with an increase in tertiary employment. Some manufacturing has remained, such as car production in Sunderland. The chemicals industry has remained important in the area throughout the period.

3 Any two suitable areas can be used.
   In the south west of the UK, especially Cornwall and Devon, the dominant industry supporting the local economy is tourism. It is often associated with sightseeing and 'bucket and spade' holidays. The nature of this tertiary employment tends to be low paid and seasonal, often semi-skilled but flexible. In contrast, areas such as Sheffield still have an important manufacturing, engineering and industrial based

workforce (based on the steel tradition). Much of this employment is technical (e.g. Boeing aerospace) and highly skilled.

76. Environmental impact of changing employment

1   Impact 1: derelict land leaves unattractive buildings. This may attract vandalism and graffiti and give a general feeling of neglect to an area.
    Impact 2: pollution makes areas look unattractive but also causes harm to wildlife, the environment and potentially people.
    Impact 3: increasing distances to market means more transport and therefore an increase in carbon in the atmosphere and air pollution as well as increasing the amount of traffic on roads, rail, air and sea.
2   Any suitable example could be used.
    For example: East London Olympic Park and surrounds. Loss of traditional industry in this part of London has helped initiate initiatives to improve the environment of the area. The London 2012 Olympic Games allowed the development of a green park, for instance, as well as supporting the Big Waterways Clean Up 2012 (BWCU2012) – a campaign to improve east London's waterways, and in particular those around the Olympic Park.
3   Any suitable example could be used.
    There were many negative environmental impacts of deindustrialisation in Sandwell in the West Midlands. Large areas of land were left derelict where houses and factories had been demolished, also parts of this land was poisoned by mercury and cadmium. The housing that was left was in a poor state for habitation and there were high levels of deprivation. Although the loss of factories had improved the quality of the air, there was still a lot of air pollution, largely due to the traffic jams because of the old and narrow roads in the area.

77. Greenfield and brownfield development

1   Greenfield sites tend to be areas of green land or countryside that has not had any previous development on it. They are often agricultural or sometimes amenity land.
2   Any suitable example can be used.
    Sheffield has a history of traditional steel making which declined hugely in the late 1970s and 1980s. It rebranded itself as the first National City of Sport, with a range of facilities originally built for the World Student Games in the city in 1991. These include the Don Valley Stadium, the Sheffield Hallam Arena and the Ponds Forge International Sports Centre, with its Olympic-sized pool and diving pool.
3   Benefits include: employment created by regeneration; many brownfield sites across the UK consist of low-demand and abandoned housing, redevelopment provides an opportunity to improve and increase the supply of new, energy-efficient homes; the government's Sustainable Communities plan of 2003 set a target for 60 per cent of new housing to be constructed on brownfield land by 2020. Other benefits include the opportunity to plant trees which encourage the use of green space for exercise (e.g. Telford in Shropshire). Disadvantages include: costs of clean-up and dealing with contaminated land before development can begin (meaning it takes longer than on greenfield sites); disturbance often means dust and water pollution.

78. New employment areas

1   There is a total of nearly 900 000 employees, and this is expected to grow to over 1.3 million by 2014. It represents nearly £100 million to the UK economy (2009 figures).
2   Any two from:
    • agriculture, especially fruit and vegetable picking
    • industrial cleaning or processing
    • health care, e.g. care workers and medicine, e.g. nurses, doctors, etc.
    • construction and building, e.g. builders, plumbers, electricians, etc.
3   See Hint. The data shows how much this sector is likely to increase in the future, but it is partly dependent on government initiatives and support, e.g. the 2013 Green Deal. Green goods, e.g. new building technologies, are also important. Some estimates put the total green sector around £5.4 billion in the UK in 2012. Led by wind and carbon finance, the UK government expects the green economy to continue to expand, growing between 4.9 per cent and 5.5 per cent a year, from 2011 to 2015. Though recent government policies have been less supportive to wind and solar power industries.

Changing settlements in the UK

79. Urban change in the UK

1   Population decline means the number of people living in an area is falling – due to emigration and low birth rate.
2   Two factors which cause population to increase are an increase in the birth rate (more people having more children) and immigration exceeding emigration, in other words more people moving into an area than moving out. The two are linked – for example, if the immigrants are young, they are likely to have children.
3   Recent government cuts have meant that some public sector workers in particular have been made redundant. This has had knock-on effects in the local economy with people spending less money.

80. Changes in urban areas

1   C
2   Most deprived areas of Derby are in a central corridor running north-south through the city. Generally, the area around the periphery at the rural-urban fringe is the least deprived, with the Mackworth Estate to the west of the centre of Derby being a notable exception.
3   'Deprivation' means an area which is lacking in things necessary for people to lead healthy lives, such as good quality housing, good access to amenities and services, plenty of jobs and educational opportunities. 'Multiple' means an area is deprived in more than one of these ways.

81. Rural settlements

1   Accessibility to large urban areas for employment and services is important so many rural settlements have developed alongside main roads – such as the overspill towns and suburbanised village shown in the diagram. Tourist villages develop in remoter rural areas so people can use them for holidays when they do not need access to the urban areas for work etc., also likely to develop close to areas of outstanding natural beauty such as National Parks. The remoter settlements have declined because people move closer to urban areas or into urban areas and therefore remote settlements lose services etc. which make even more people leave.
2   Any one from: improved speed and availability of transport – especially cars and trains – mean people can still commute to their work in urban areas; more people now work for themselves from home – increase in home-working generally because of new technology, such as the internet; changes in lifestyles and wealth mean people want bigger houses and gardens – these are more available and less expensive than in urban areas.
3   Any suitable examples could be used: south coast, Brighton, Eastbourne, Devon and Cornwall. These places have attracted retired people because house prices are cheaper than the urban areas they have come from (especially London), so selling their London homes increases their pension; climate is mild; coastal towns are attractive; calm and quiet of coastal towns appeals to older people rather than the young who want nightlife etc.; there are fewer jobs in these areas than in urban areas but that does not matter for retired people – younger people need the job opportunities presented by large cities such as London; these areas have developed to meet the needs of older people – good care homes; transport; health care, etc.

82. Contrasting rural areas

1   A or B
2   Any two contrasting areas could be used, but one should be a remote rural area and the other an accessible rural area. Accessible countryside, e.g. north Wiltshire, has low unemployment and a range of employment whereas a remote rural area, such as the Scottish Highlands, has a far more limited range of employment opportunities – mostly primary industries such as farming, fishing and forestry. Economy of the Highlands is therefore almost totally dependent on agriculture and tourism – makes it very vulnerable, for example, when seasons are poor. Access to services is very different – in north Wiltshire there is good access to medical, educational and retail services, though most are car dependent. In contrast, the service provision in the Highlands is poor – people have to travel a long way,

usually by car to access hospitals, schools etc. which can make life difficult. North Wiltshire's population is more varied and population is increasing which is putting a strain on some services and housing provision. The Highlands, especially in some areas, has an ageing population and associated problems as young people leave for work in more urban areas. However, remember that this does not mean that quality of life in north Wiltshire is much better than in the Highlands – living in one of the most beautiful parts of the UK has many benefits for leisure and recreation.

83. Impact of housing demand
1   Any one from: building on people's gardens or other green spaces in towns and cities means the environment is less pleasant for people, will not attract as many animals and birds etc.; pressures on green belt and areas at rural-urban fringe; new housing means more waste and more pollution – both in living and building them; increases traffic which increases smog and pollution.
2   Any suitable examples can be used. Since the 1980s, the London Docklands areas has been transformed through regeneration and rebranding – new housing has been built (including some low-cost homes) and older housing has been improved; new services such as health, education, leisure and retail services have been created; transport links into and within the area have been improved – Docklands Light Railway, city airport etc. Canary Wharf has become an important business area which provides many jobs. All of these have improved housing, services and the economy. However, not everyone likes the changes and believes they do not benefit all people, such as unskilled workers.
3   Rebranding an area means giving it a new image so that it will attract businesses and development. Rebranding needs to be done in accordance with practical action for it to work. See answer to question 2 about London Docklands and Canary Wharf. Other examples could include Castle Wharfe, Nottingham: old canal area was derelict. Initial work by the City Council improved the site – landscaped, planted and paved. Some businesses relocated there (Magistrates Court, Inland Revenue, County Archive), bars, banks, offices (BT, Evening Post) also moved in. Transport to the area also improved and pleasant environment has encouraged apartments and more offices to be developed there. Scarborough is an example of a not very successful rebranding exercise. In 2004, £2.8 million invested in improving the harbour area but the area is still dominated by cheap shops, cafes, amusement arcades and the water quality at the beach is not the best so tourist attractions are still limited.

84. Making rural areas sustainable
1   In National Parks, all development is very strictly controlled. Similarly, green belt areas have strict planning controls.
2   See Hint. Policies to conserve the landscape include green belts and National Parks. Both types of policy have had mixed success. Green belts have slowed down urban sprawl around cities and towns and restricted development; National Parks have preserved beautiful landscapes. Both green belts and National Parks are enjoyed by locals and tourists. They also provide employment for locals, for example quarrying in the Peak, Yorkshire and Snowdonia national parks. However, some areas of green belt have been developed in spite of planning rules. Visitors to National Parks and green belts can damage the environments, e.g. overuse of footpaths, litter, etc. There's a lack of affordable rural homes, which cannot be built in National Parks or on most green belt land. It is sometimes difficult to build house extensions or set up a business.
3   See Guided. Also, the EU's Convergence Objective invests in rural businesses such as the Eden Project which now provides 400 jobs and attracts over a million visitors every year, boosting the local economy.

The challenges of an urban world
85. Global trends in urbanisation
1   B
2   Between 1990 and 2025, China's population will rise from around 1150 million in 1990 to an estimated 1400 million in 2025. The rise between 1990 and 2011 was greater than the rise predicted between 2011 and 2025.
3   Any one reason from: increased employment opportunities in urban areas; fewer employment opportunities in rural areas; better wages in urban areas; better access to services in urban areas; people moving to urban areas to be with family and friends who have already moved.

4   An increasing proportion of a country's population living in urban areas – towns and cities – rather than in the countryside.

86. Megacities
1   C
2   Most megacities are in developing countries with the exception of some USA and Japanese cities and Seoul – this is because many developing countries have very large populations (e.g. China and India) and increasing urbanisation. Within the developing world, there is a concentration of megacities in south and east Asia, where populations are very large.
3   Rural-urban migration in most developing countries – especially in South-East and East Asia as people move to the cities looking for better employment and educational opportunities. These countries also have growing populations because of natural increase – a high birth rate meaning the population is increasing. Many young people move to the cities and this is the age group which has children.

87. Urban challenges: developed world
1   Any one from: waste disposal – it is expensive and difficult to hygienically dispose of waste in such a large urban area; transport – to try to reduce congestion and pollution of transport and help people get around more easily; energy use for such a city in the developed world is huge – need to try and cut down to reduce emissions and pollution.
2   The port suggests that this city has a lot of trade which therefore attracts businesses which brings people to live and work in the city; high rise buildings show there is lots of tertiary and quaternary industry which would also attract people to the area; living close to the river would help with the city's water requirements.
3   Any of the following are being tried to reduce traffic congestion and pollution: congestion charging – charging people for using cars in certain areas, therefore encouraging the use of public transport, walking, cycling, etc.; setting up car share schemes; providing cheap, efficient public transport; creating pedestrianised zones so no traffic is allowed in certain parts of an urban area except for deliveries to businesses etc.

88. Urban challenges: developing world
1   The informal economy is business or employment that is not officially recognised – people work for cash, often for themselves on the streets.
2   Any examples will be credited here. Any one from: limiting / reducing traffic congestion to limit air pollution from exhaust fumes – e.g. congestion charging, pedestrianisation, cycle lanes, cheap and efficient public transport; setting laws and / or charging businesses for the amount of air pollution caused by factories etc.; providing good waste disposal so rubbish does not end up polluting rivers / the water supply; providing good sanitation to prevent pollution of water supply.
3   There are some similarities – both developed and developing world cities face the challenge of dealing with pollution and traffic congestion, though the pollution may be of different types. Also, both face problems of overcrowding but to very different extents and with different consequences. Developing cities such as Sao Paulo in Brazil have shanty towns which creates a lot of problems – lack of electricity, access to clean water, no rubbish collection or waste disposal which all lead to pollution and poor living conditions, causing disease and death. Developing cities also face the challenges of more employment in the informal sector where pay is low and there is no job security. People working in the informal sector are also more likely to experience exploitation and mistreatment. Developed cities usually have more waste per person to dispose of and usually use more energy for things such as household appliances, which few people in the developing world can afford.

89. Reducing eco-footprints
1   The eco-footprint is a measure of the impact of a place (or person) on the environment. It is explained in terms of how much land per person is needed to support the city's use of energy, water, food and waste against the pollution generated.
2   Different places have different eco-footprints because of the different lifestyles and facilities in place. For example, if a place has many things in place to dissuade people from using their cars, such as an efficient and sustainable public transport

system, then that place will have a smaller eco-footprint than somewhere where there is very high car use because people are encouraged or not dissuaded to use cars.

3 Any suitable city could be chosen. For York, the local council distributes leaflets and advice on how to reduce energy use such as by turning down the heating. It encourages people not to use their cars, which reduces pollution as well as reduces energy use, through a park and ride scheme, bicycle lanes, a city car club, car-sharing scheme and a pedestrian zone where traffic is not allowed. To try and reduce its waste, the council in York provide a good recycling service for households and businesses. It also recovers over 70 per cent of the methane from landfill sites and uses it for energy.

90. Strategies in the developing world

1 See Guided. Pollution is also a big social and environmental problem. Traffic congestion and poorly maintained cars generate serious air pollution, creating health problems. Dense smog can cover the whole area of Mexico City. Rivers and seas are used as dustbins, destroying wildlife. As the cities spread outwards, wildlife habitats are destroyed as well as agricultural land. Supplies of underground water beneath some cities are being used up so fast that the land on the surface is sinking. Mexico City has sunk 7 m in the last 100 years. Any suitable example will be credited.

2 Any suitable example will be credited. Curitiba was awarded the Global Sustainable City award in 2010. It has several measures which improve people's quality of life. It has preserved green spaces with 28 parks and wooded areas which people can enjoy; people swap bags of their rubbish for bus tickets and food; children can exchange recycled rubbish for school supplies and food; the fast, cheap and efficient transport systems helps to transport 2.6 million people each day with minimal pollution. Other examples include work done by NGOs such as Urban Green Partnership Programme which created 300 home gardens in Sri Lanka so people could grow more of their own food and have more green space to enjoy. Centre for Urban and Regional Excellence has provided a simple waste water and sewage plant to prevent the pollution of water supplies in Kachpura, in India, so fewer people are becoming ill.

The challenges of a rural world

91. Rural economies

1 Most subsistence farming is south of the equator with some just north of it. No high income countries have subsistence farming. It is mostly concentrated in Africa and South and South-East Asia with some also in South and Central America.

2 D

3 Any one from: quality of the farmland cannot support anything other than subsistence farming; people are too poor to afford the machinery and equipment necessary to produce more from the land; pieces of land held are too small for more to be produced; people lack the education necessary to improve the land through things such as irrigation schemes which would improve fertility; distances / access to markets to sell produce is too difficult to make it worth growing more.

92. Rural challenges: developed world

1 D

2 Generally the larger the parish population, the better is the access to services – or the smaller the parish, the fewer services are available. For the smallest parishes of up to 1000 people, 48 per cent have no post office and 49 per cent have no shop – this goes down for parishes of between 1000 and 3000 people, and goes down again so parishes of over 3000 people just 4 per cent have no post office and 10 per cent have no shop. The only statistic which slightly bucks the trend of services diminishing the smaller the population is, is with those parishes which have no bus service at any time – this applies to 9 per cent of parishes over 3000 but 8 per cent of parishes of 1000–3000.

3 Tourists may bring food and supplies with them and therefore not use local services – meaning local services may close; tourist accommodation and especially 2nd homes used as holiday homes can cause house prices to rise so local people cannot afford to live there; presence of a large number of tourist houses / accommodation means the rural area is very busy during some times of the years; seasonality means lack of employment for local people during some times of the year; providing services for tourists may come at the expense of services to locals; tourists can damage the environment – large, unsightly hotels, producing a lot of

waste and using a lot of energy, causing pollution and damage to wildlife, etc.

93. Rural challenges: developing world

1 Any suitable example could be used. In Kenya, for example, 50 per cent of the people are classed as poor and most of these – 75 per cent – live in rural areas. Some farmland is very rich and fertile but most of this is taken by huge commercial farms, which export their produce and therefore have little benefit for many Kenyan people. Of the land left, most of it is used for subsistence – families using the land to support animals and grow crops just to feed themselves with little over to sell. Yields are low because of lack of education, meaning farming techniques are poor and land degradation is occurring. Rural-urban migration of young men means that the majority of rural farm workers are old and / or female. Young men migrate to the towns and cities in search of work as there are very few rural paid jobs available. There is also a high rate of HIV and AIDS in the rural population, making people too ill to farm and needing to be looked after by others who are therefore also unable to farm.

2 Specific suitable examples should be included. You could use a number of examples from one place or examples from different places. Some rural areas of the developing world are places where frequent natural hazards occur, such as floods, droughts and earthquakes which kill and injure people as well as damaging the land and crops that grow on it. Human hazards such as war can also badly damage the land and rural settlements. Diseases such as HIV / AIDS, malaria, cholera, TB, etc. all increase the death rate, meaning a greater need for health care as well as taking members of the population away from paid work. Population growth in developing countries puts pressure on land to produce more food and resources to resource the population – can damage the land and have the opposite affect. International tourism and globalisation means that many large businesses, hotels etc. are not owned or run by locals so few benefit – money is take out of the country.

94. Rural development projects

1 Any suitable examples should be used. In Koraro, in Ethiopia, the NGO Millennium Promise has provided three micro dams and 30 safe water points – improves access to water for irrigation etc. and ensures this water is clean so will not harm people or the environment. Has also provided school resources to improve education for the young, and malaria nets to help prevent malaria. Also runs training programmes for local farmers, which have improved crop yields and therefore farmers' incomes. In Afar, in Ethiopia, Farm Africa has built irrigation systems to improve access to water in an area often affected by drought, given farmers seeds that they would not be able to afford so they can feed themselves and have some produce to sell, given farmers loans to invest in farm tools or crops to improve their farms and get higher yields.

2 Various different groups try to improve rural life. For example, NGOS – charities such as those given in answer to question 1 which set up and run development projects in rural areas. The United Nations and other intergovernmental organisations provide rural areas with money to improve – this is donated by individuals and governments. Developing country's government uses tax money to fund development projects – often on a large scale. Local government / councils of rural areas fund smaller-scale projects. Local people and communities work together to raise money for their own small-scale projects.

95. Developed world: farming

1 Any one from: traditional farming no longer provides them with enough income; diversification may be less hard work than traditional farming to raise money; diversifying means a farm's income is spread over a wider area – therefore makes sure that farms can still earn money even when, for example, they cannot sell livestock because of TB, milk prices are lower than the costs of producing milk, crops may fail, etc.

2 Any specific suitable examples could be used: providing accommodation for tourists, e.g. B&B, camp sites, holiday cottages; using land for leisure activities such as mountain biking, shooting etc.; branching out into new products such as alpaca and ostriches, producing their own products such as cheese, butter, yoghurt, etc. and selling them in their own farm shop and / or at farmers' markets; pick your own fruit; converting farm buildings into business premises for people to let or buy, leasing the land for wind turbines, solar panels etc.; selling farmland / buildings for housing.

96. Developing world: farming

1 Any two from: farmers are paid an agreed minimum price all the time so it is not entirely subject to market forces and so increases income for farmers; farmers have to invest some money in local community projects which they benefit from as well; working conditions have improved as they are inspected; inspections make sure the environment is not being damaged so the farm should continue to be environmentally and economically sustainable.

2 Any one from: equipment / tools are not expensive so local people can afford them; they are not difficult to use to do not require much training; they are easy and cheap to maintain and repair; usually help improve environmental sustainability.

3 In Zimbabwe, for example, small petrol and diesel powered pumps give farmers who have no access to electricity an efficient means of lifting water for crop irrigation; digging boreholes means water deep underground can be accessed; sewage water can be recycled in shallow lagoons; barrels can store rainwater for use later; small dams can be built on rivers.

UNIT 3: MAKING GEOGRAPHICAL DECISIONS

99. Information on the problem

1 B

2 D

3 Haiti is an island in the Caribbean, between Cuba and the Dominican Republic.

4 Development is one of a series of measures which relates to the standard of living and quality of life for a country's population.

5 See Guided. Other points may include the following.
- There may also be little money for preparedness systems to protect people, such as active monitoring and evacuation procedures (e.g. tropical storms).
- Low levels of development mean that it will also be difficult for a population to easily recover following a natural hazard event because there is little reserve budget for recovery.

101. Different levels of development in Haiti

1 Haiti has a low level of development, similar to countries that are part of Sub-Saharan Africa.

2 Haiti has a lower life expectancy than the rest of the Americas. This is undoubtedly due to the low levels of development and the problems that result, such as poor education, lack of fresh water, ill health, diseases, diet etc.

3 Key factors might include:
- urbanisation and a rapidly growing urban population
- poor sanitation and water supplies
- frequency of natural hazards (multiple exposure to landslides, hurricanes and earthquakes) are also important.

Try to describe the factor and then explain why it contributes to low life expectancy.

103. Impacts and costs of the disaster

1 D

2 1 Seismologist 1 notes that lots of buildings were damaged.
 2 An earthquake of magnitude 7.3 close to the capital is a very strong earthquake in a country with low levels of development.

3 The most intense shaking centred around the capital, Port-au-Prince. Shockwaves seem to extend up to 100 km away (Figure 3a). Figure 3c shows a line eastwards (red area) from the capital.

4 See Guided. You can also link this answer to what you have studied on development. For example, the Haitian population lived in poorly constructed buildings (Figure 1b) that subsequently collapsed, leading to a very large death toll (300 000 people, Figure 3b).

105. Aid and development

1 Initially money will need to be spent on rebuilding infrastructure and housing but this need will decrease over time as developments are completed (a fall from estimated $2 912 million in 2010 to $444 in 2013).

2 C

3 Recovery costs are often associated with dealing with the immediate after effects of the disaster and include medical treatments, emergency shelters, etc. Reconstruction costs, however, tend to be more long term so they may include infrastructure, e.g. schools and hospitals.

4 The cartoon shows the fact that the earthquake itself acts as a barrier to relief / aid (which may have been happening in the country even before the event itself). Following the earthquake, further aid cannot get into the country because the infrastructure and port have been destroyed. As a result, it cannot be distributed to the people and there are many deaths.

107. Haiti's water and education status

1 See Guided. Figure 5a shows that the number of sick or dying from cholera is increasing daily (1526 to 4147 sick in just 5 days). Cholera is a waterborne disease, so distributing clean water is a priority in order to reduce the incidence of infection and help the people of Haiti recover from the earthquake. Other points may include: there is also increasing pressure of population, especially in urban areas, which is putting additional strain on the fresh water supplies and foul water drainage systems. Fresh water is essential to the functioning of all modern cities.

2 Points may include: there are a number of significant challenges which are linked to high rates of population growth and a rapidly growing urban population especially in the capital, Port-au-Prince. This puts strains on the country's education system and may lead to large class sizes. There may also be pressure to go out and work, rather than stay in education. Haiti is ranked 177th out of 186 countries in the world for education spending, therefore, a significant challenge is to redirect a greater proportion of the country's limited GDP into primary education. The World Bank has reported that Haiti needs a free, public and universal school system. If education levels were higher, levels of development may be improved as literacy levels, employment prospects and skills-base would be improved.

3 See Guided. If less rain falls, the dry season will get longer and put a strain on existing water supplies. Water is needed not only for drinking but for farming too, so a reduction in water supplies will reduce crop yields and livestock.

108–109. Making decisions

1 Try to make up to three developed points, giving examples from the sources and your other knowledge from Units 1 and 2 to support them. Further points might include: this would develop basic skills of reading and writing, which in the long run would be beneficial to the country's economy and help with growth and the longer-term recovery which the country desperately needs. Education may also help in the prediction and management of natural hazards as better education goes hand in hand with community preparedness – a clear advantage. Improving primary education would increase the chances of more people from Haiti going on to further education and getting better paid jobs. This could become a disadvantage to Haiti in the long run, encouraging more people to leave the country in search of good jobs and a better quality of life. If they leave, they will not invest their money in Haiti and the economy will not improve, meaning further development will not occur.

2 You should include arguments for your choice and justify why you feel it is the best choice. You may also recognise the merits of the other options. Try to make three or four well developed points, using evidence from the sources and from your knowledge of Units 1 and 2. Further points might include: there may also be an increase in the frequency of storm hazards, again pressurising this precious resource. Although the other options are undoubtedly important, aid money must be focused on developing better clean water supplies, which are essential for life and basic livelihoods. The cholera episode that occurred in October 2010 was as a direct result of poor water infrastructure. Water.org have put in place ideas for the development of improved water infrastructure (50 000 people – Figure 5b) this simply is too small scale compared to the population of the island (7 million) and the capital (700 000 people). Along with improving water and sanitation, education should be improved with more of the population being given the opportunity to receive a free primary education. In this way, people will better understand the importance of good hygiene practices – not drinking contaminated water and preventing water becoming polluted in the first place. With access to education, the people of Haiti can learn to save water and maintain water sources such as wells and the use and repair of pumps.

Published by Pearson Education Limited, Edinburgh Gate, Harlow, Essex, CM20 2JE.

www.pearsonschoolsandfecolleges.co.uk

Copies of official specifications for all Edexcel qualifications may be found on the Edexcel website: www.edexcel.com

Text and original illustrations © Pearson Education Limited 2013
Edited, produced and typeset by Wearset Ltd, Boldon, Tyne and Wear
Illustrated by Wearset Ltd, Boldon, Tyne and Wear
Cover illustration by Miriam Sturdee

The rights of David Holmes to be identified as author of this work have been asserted by him in accordance with the Copyright, Designs and Patents Act 1988.

First published 2013

17 16 15 14
10 9 8 7 6 5 4 3 2

British Library Cataloguing in Publication Data
A catalogue record for this book is available from the British Library

ISBN 978 1 446 90538 8

Acknowledgements
The publisher would like to thank the following for their kind permission to reproduce their photographs:

(Key: b-bottom; c-centre; l-left; r-right; t-top)

Alamy Images: charistoone-stock 29, DWImages Northern Ireland 82, Finnbarr Webster 87b, geogphotos 25, Justin Kase 95, Richard Wareham Vervoer 60, The National Trust Photolibrary 33; **Co-operative Group:** 96; **David Holmes - Geography Education:** 3; **DK Images:** Alex Havret 55; **Getty Images:** The Image Bank / Niels Busch 45l; **Masterfile UK Ltd:** Jeremy Woodhouse 87t; **National Trust Photo Library:** Mark Sunderland 16; **Photos.com:** 9, Ronald Hudson 10; **Press Association Images:** AP / Dario Lopez-Mills 97; **Rex Features: Robert Harding World Imagery:** Still Pictures / Mark Edwards 106; **Shutterstock.com:** Fedor Selivanov 26, Konstantin Shevtsov 45r, think4photop 38; **www.CartoonStock.com:** Brian Fray 68

All other images © Pearson Education Limited

We are grateful to the following for permission to reproduce copyright material:
Figures
Figure on page 39 from http://www.unep.org/dewa/vitalwater/article177.html Map attributed to R J Diaz and R Rosenberg 1995, With kind permission from Professor Robert J. Diaz; Figure on page 61 from http://www.geographyinthenews.rgs.org/resources/images/Clark-Fisher_news_en.gif, Royal Geographical Society, Royal Geographical Society (with IBG), www.geographyinthenews.rgs.org ; Figure on page 62 fromhttp://www.fao.org/gender/infographic/en/ (Click the first box Why are women so important to agriculture?) Food and Agriculture Organization of The United Nations, Food and Agriculture Organization of The United Nations (2013) Why are women so important to agriculture? http://www.fao.org/gender/infographic/en/ Reproduced with permission; Figure on page 100 from gapminder.org, Data from UNESCO Institute for Statistics; Figure on page 100 from gapminder.org, Data from Human Mortality Database http://www.mortality.org/ Human Lifetable Database: http://www.lifetable.de/ UN World Population Prospects: http://esa.un.org/wpp/

Maps
Map on page 64 from http://3.bp.blogspot.com/_q016kExGrVs/TReSLlwH-6I/AAAAAAAAAic/1lmHd6s2QbU/s1600/routes.png Original source: http://openflights.org/data.html; Map on page 80 from https://www.gov.uk/government/publications/english-indices-of-deprivation-2010, Crown Copyright / Open Government Licence; Map on page 102 from http://www.cbc.ca/gfx/images/news/photos/2010/01/13/haiti-shakemaplegend-584.jpg, CBC.ca, US Geological Survey

Tables
Table on page 42 from http://www.ospar.org/documents/dbase/publications/p00618/p00618_2012_mpa_status%20report.pdf Data from 2012 Status Report on the OSPAR Network of Marine Protected Areas, © OSPAR Commission, 2013; Table on page 74 adapted from Office of National Statistics Annual Abstract of Statistics and Office of National Statistics United Kingdom National Accounts, Source: Office for National Statistics licensed under the Open Government Licence v.1.0.; Table on page 79 adapted from https://www.gov.uk/government/uploads/system/uploads/attachment_data/file/224068/bis-13-p143-low-carbon-and-environmental-goods-and-services-report-2011-12.pdf, Contains public sector information licensed under the Open Government License v1.0; Table on page 92 fromhttp://www.poverty.org.uk/71/index.shtml Original data from the Rural Services Survey 2000, / DEFRA, With kind permission from The Poverty Site, www.poverty.org.uk; Table on page 98 from CIA World Factbook Haiti, https://www.cia.gov/library/publications/the-world-factbook/geos/ha.html

Text
Extract on page 28 adapted from http://www.environment-agency.gov.uk/homeandleisure/107550.aspx, Environment Agency 2012; Extract on page 35 adapted from 'Will climate change lead to more flooding?', *The Guardian* (Duncan Clark), http://www.theguardian.com/environment/2012/oct/08/climate-change-more-floods Grantham Institute, Imperial College London and Duncan Clark the guardian.com, Monday 8 October 2012 09.31 BST, With permission of Guardian News and Media Ltd; Extract on page 79 adapted from Davis Hill's London Blog :London's population up by 12% in 10 years, *The Guardian* (David Hill), http://www.theguardian.com/uk/davehillblog/2012/jul/16/london-population-rises-12, With permission of Guardian News and Media Ltd

Every effort has been made to trace the copyright holders and we apologise in advance for any unintentional omissions. We would be pleased to insert the appropriate acknowledgement in any subsequent edition of this publication.

In the writing of this book, no Edexcel examiners authored sections relevant to examination papers for which they have responsibility.